Organizing for a Complex World

Significant Issues Series
Timely books presenting current CSIS research and analysis of interest to the
academic, business, government, and policy communities.
Managing Editor: Roberta Howard Fauriol

About CSIS
In an era of ever-changing global opportunities and challenges, the Center for
Strategic and International Studies (CSIS) provides strategic insights and practical
policy solutions to decisionmakers. CSIS conducts research and analysis and develops policy initiatives that look into the future and anticipate change.

Founded by David M. Abshire and Admiral Arleigh Burke at the height of the
Cold War, CSIS was dedicated to the simple but urgent goal of finding ways for
America to survive as a nation and prosper as a people. Since 1962, CSIS has
grown to become one of the world's preeminent public policy institutions.

Today, CSIS is a bipartisan, nonprofit organization headquartered in Washington, D.C. More than 220 full-time staff and a large network of affiliated scholars
focus their expertise on defense and security; on the world's regions and the
unique challenges inherent to them; and on the issues that know no boundary in
an increasingly connected world.

Former U.S. senator Sam Nunn became chairman of the CSIS Board of Trustees
in 1999, and John J. Hamre has led CSIS as its president and chief executive officer
since 2000.

CSIS does not take specific policy positions; accordingly, all views expressed
herein should be understood to be solely those of the author(s).

The CSIS Press
Center for Strategic and International Studies
1800 K Street, N.W., Washington, D.C. 20006
Tel: (202) 775-3119   Fax: (202) 775-3199
E-mail: books@csis.org   Web: www.csis.org

Organizing for a Complex World

Developing Tomorrow's Defense and Net-Centric Systems

Edited by Guy Ben-Ari and Pierre A. Chao

Foreword by David J. Berteau

THE CSIS PRESS

Center for Strategic and International Studies
Washington, D.C.

Significant Issues Series, Volume 31, Number 1
© 2009 by Center for Strategic and International Studies
Washington, D.C.
All rights reserved
Printed on recycled paper in the United States of America
Cover design by Robert L. Wiser, Silver Spring, Md.
Cover photograph: © HIROYUKI OTSU/amanaimages/Corbis

13 12 11 10 09   5 4 3 2 1

ISSN 0736-7136
ISBN 978-0-89206-551-6

Library of Congress Cataloging-in-publication Data
Organizing for a complex world : developing tomorrow's defense and net-centric systems / edited by Guy Ben-Ari and Pierre A. Chao.
    p. cm. — (Significant issues series ; v. 31, no. 1)
    Includes bibliographical references and index.
    ISBN 978-0-89206-551-6 (pbk. : alk. paper) 1. United States—Defenses. 2. United States. Dept. of Defense—Management. 3. United States—Armed Forces—Technological innovations. 4. Defense industries—Technological innovations—United States. 5. Weapons systems—Technological innovations—United States. 6. Systems engineering—Technological innovations—United States. 7. Information technology—Government policy—United States. 8. Computer networks—Government policy—United States. 9. Computational complexity. I. Ben-Ari, Guy, 1973– II. Chao, Pierre A. III. Title. IV. Series.

UA23.O69 2009
355'.070973—dc22   2008043680

CONTENTS

Foreword · *David J. Berteau*   vii

Preface · *Pierre A. Chao*   xiii

Acknowledgments   xv

1. Introduction: Framing the Complexity Challenge · *Guy Ben-Ari and Matthew Zlatnik*   1

2. Making Governance Matter More: Oversight, Insight, and Foresight in Complex Systems Procurements · *Michael Schrage*   10

3. Models for Governing Large Systems Projects · *Harvey M. Sapolsky*   24

4. Competition and Innovation under Complexity · *Jeffrey A. Drezner*   31

5. Systems Integration for Complex Defense Projects · *Eugene Gholz*   50

6. A New Way of Thinking about Enterprise Capability Development: Network-Centric, Enterprise-Wide System-of-Systems Engineering · *Jeremy M. Kaplan*   66

7. Engineering of Complex Systems: Challenges in the Theory and Practice · *Douglas O. Norman*   82

8 Managing Megaprojects: Lessons for Future Combat Systems · *Marco Iansiti*   88

9 Managing Government Effectively in a Complex Environment: Influencing Networks through the Network Campaign · *W. Scott Gould and Julie M. Anderson*   111

10 Human Capital for Complexity · *David H. Dombkins*   128

About the Editors and Authors   159

Index   165

FOREWORD

DAVID J. BERTEAU

Moore's Law has fueled advances in computing since it was first proposed 40 years ago by the cofounder of Intel. The computer chip manufacturer built its business plan on the basic premise that the computing power of microchips will double every 18 months. Although not a "law" in the scientific sense of the word, Moore's Law has largely held true for semiconductors for decades, which has been good for business for computer chip and computer makers. It has been even better for those whose challenge is to manage large amounts of information. Thanks to the law, the storage of data has become cheaper and cheaper each year; today, a student can carry the equivalent of 20 Encyclopedia Britannicas in a flash drive the size of a toddler's thumb and can search through them with a device the size of a deck of playing cards.

It has long been a truism that we cannot manage what we do not know. The expansion of data storage, access, and manipulation should, it seems, enable managers to tackle tasks of increasing complexity with success. However, experience and analysis show that better information is not enough; success is engendered or undermined by many elements beyond data management. Thus far, no reliable research has conclusively demonstrated a linear cause-and-effect relationship between better data management and greater success in the development and production of complex systems.

Nowhere is this clearer than in the business of building defense systems. These systems—ships, tanks, aircraft, satellites, et cetera—are generally the most complex systems undertaken today, incorporating technology not yet developed when the systems are being designed. They are actualized by a team of government managers and industry practitioners, aided by a vast assemblage of engineering and scientific talent, overseen by political forces, monitored by auditors at every step, regulated by rules measured in linear feet, and ultimately evaluated in life-and-death situations. This is tough work.

Two decades ago, David Packard's commission on defense management took a hard look at this complexity and tried to compare it to programs of equivalent complexity in the commercial and other government sectors. The Packard Commission concluded in 1986 that "defense is different"—in scope, complexity, and consequences. Few parallels, and none outside of government, could be found.

Yet, what Packard discerned was not new. The challenge of managing the design and construction of complex defense systems has been with the United States since its beginning. As Ian Toll so captivatingly describes in his 2006 book *Six Frigates*, the first challenge of the early U.S. Navy was managing a complex system, which was the design and construction of its first six frigates. In an eerie precursor to today's security environment, the task called for responding to one threat that was active and visible, pirates and corsairs from the Barbary Coast, while preparing for an entirely different one that was far off and not widely perceived, war with a major European power like France or Britain. Questions that occupy analysts today were present in 1794: whether to have the prime contractor make or buy parts, whether to convert from commercial variants to military use or to design from scratch, whether to meet schedule or enhance performance, how to explain cost overruns to Congress. The response was a new design, untested and largely untestable. The result was a breakthrough in naval weaponry that enabled the new nation to survive, taking advantage of a largely unrecognized gap in capability between frigates and ships of the line.

These are the stakes of complex systems in the Department of Defense. The success or failure of military operations depends on them. The lives of America's sons and daughters hang in the balance. The defense of national interests demands success. That is why this book is so timely and worth reading.

Consider the recent record.

The Defense Science Board's task force on integrating commercial systems reviewed several systems that, on paper, should have been easy to develop and deliver on schedule—the Navy's Littoral Combat Ship, the Army's Armed Reconnaissance Helicopter, and the replacement for the presidential helicopter known as Marine One. It found that, at the time of contract award, each system's design was based on a commercial variant and that militarizing a base platform would save time and money over building a new system from scratch. In each case, those savings have not materialized, and the task force found many causes for that outcome. In each case, though, managers for the government, the prime contractors, and the commercial subcontractors shared one common feature: they underestimated the complexity of requirements, integration of subsystems, and the interaction of changes in one subsystem with new demands on others.

One reason for these shortcomings in management is that the federal government's capability and capacity for systems integration has declined over the last two decades. At the height of the Cold War, defense systems commands (such as the Naval Air Systems Command or the Air Force Systems Command) had the workforce needed to manage complex systems, using a combination of military, civilian, and outside personnel. Assistance for systems-of-systems integration was the purview of research centers and government labs.

Since the 1990s, and continuing after 9/11, this capability has been diminishing. In review after review, experts conclude that the Defense Department will probably need a substantial increase in its capability to integrate systems of systems. How will that capability be created? What form of governance will be needed to sustain this increase over time? These are key questions that the contributors to this volume begin to address.

Faced with diminished internal government capability, the Department of Defense and other national security agencies like the Coast Guard and the Customs and Border Protection Agency have moved to reliance on prime contractors as Lead Systems Integrators, or LSIs. Congress and audit agencies like the Government Accountability Office (GAO) have found it easy to criticize their performance, but this criticism bears a second look.

All too often, contractors are blamed for situations that the government is partially responsible for having created. This is nearly always the case for problems in major system cost, schedule, and performance. Both the government and its contractors participate in fundamental causes of these problems, and these causes merit some attention, because no serious effort can be made to improve management of complexity without addressing the underlying challenges.

First is the need to define requirements better. In its final report in early 2006, the Defense Acquisition Performance Assessment project concluded that inadequate requirements were a core cause of schedule delays and cost overruns in major defense systems. At the other end of the contracting spectrum, the Gansler Commission (officially, the Secretary of the Army Commission on Acquisition and Program Management in Expeditionary Operations) found in late 2007 that similar requirements inadequacies were one major cause of contracting problems in support of expeditionary operations in Iraq and Afghanistan. Obtaining clear requirements is a prerequisite to contracting success in complex systems as well as in all government contracting.

In addition, formal system requirements are difficult to change. The process of reaching a decision through the Joint Requirements Oversight Council in the Defense Department can take two years or more, making any system manager reluctant to raise questions that could cause that process to be restarted. Yet, requirements should permit users and developers to be smarter today than they were yesterday. System design goals should be adjusted accordingly. Today, that rarely happens.

Second is the need to improve the front end of the process for contracting. This pre-award process is the government's way of converting requirements into a solicitation document, then seeking bids from potential contractors. The process includes the scope of work that will be performed by the winning bidders and the criteria for evaluating their bids. It also includes the government's evaluation of those bids and selection of the winner or winners.

In recent years, the results of this pre-award process have been less successful than in the past, as measured by the number of successful protests lodged by losing bidders with GAO. In most of the successful protest cases, loser positions are upheld because of procedural flaws by the government in solicitation, evaluation, and award of contracts. In some cases, good decisions have been undermined by governance failures as simple as inadequate documentation. Process failures need to be reversed as a minimum condition of success in organizing for complex systems.

Third is the need to improve the management of contracts following award. For complex systems, this post-award process is hard enough even with clear requirements and a pristine pre-award process, because the tasks under contract are challenging and difficult to achieve. Yet the quality and quantity of post-award personnel, the contract administration organizations, have been dramatically reduced since 1990, and the process of restoring them has yet to begin.

Underlying all of these root causes is an overall policy framework that does not make clear the roles and functions of the varying participants. The government, its R&D centers and labs and supporting contractors, and the contractors who deliver systems operate under a set of policies that make it difficult to tell where the work of one participant should end and another begin. Compliance with these loosely defined policies is impossible to measure.

The problems listed here permeate the entire universe of government contracting but are particularly important for the management of large, complex systems. Fixing problems there will require new thinking by both the government and industry—particularly in making trade-offs and in defining and measuring success.

Ultimately, the value of a systems-of-systems integrator is to help the Defense Department make trade-off decisions. Providing that help requires broad access to knowledge across all potentially applicable systems and subsystems and all components and specialties. Access to knowledge is something that can, in fact, be measured. The number of systems and subsystems and components and specialties are known (or at least knowable), and knowledge of them can be documented and measured with considerable precision.

In one possible approach, stable teams of talented scientists and engineers can be assessed against all the access-to-knowledge categories (systems, subsystems, components, technologies). Those measures can be both *relative* (that is, comparing Defense Department labs, R&D centers, and private contractors) and *absolute* (do we have enough? Is everything covered?).

Perhaps equally important, if the clash of ideas and the evaluation of trade-offs among those ideas really lead to better solutions, then the measures above may also support an assessment of who can do a better job of systems-of-systems integration by fostering that clash of ideas and that trade-off capability. Such a process could move the issue from one of emotion and philosophy to one of analysis and metrics.

What may be needed therefore is a way to tie the underlying Department of Defense skill base to systems-of-systems integration by connecting it to access to knowledge, both current and emerging. In so doing, we may be able to address both our concerns—how do we organize for better management of complex systems, and how do we measure success?

This path will not be easy, nor will progress be rapid. The assessments offered in this volume, however, provide a good start along that path. They also provide the much-needed balance of competing viewpoints, offering alternative prisms through which to view the reality that we have yet to define clearly. The Defense-Industrial Initiatives Group at the Center for Strategic and International Studies is pleased to have played a role in beginning this stage of the journey toward better management of complex systems. As we continue to work toward this goal, we at CSIS welcome participation by all who read this book.

PREFACE

PIERRE A. CHAO

The genesis of this volume was a set of comments by Deputy Secretary of Defense Gordon England and Undersecretary of Defense for Acquisition, Technology, and Logistics Ken Krieg in early 2006. Both remarked that one of the most significant challenges facing the Pentagon was the ability to manage the rising complexity of its weapons and defense systems programs.

Historically, there has always been a tension between the increasing complexity of tasks being taken on and the technology and management tools available. Whether it was the development of the Gantt chart used to build the Navy ships of World War I or the development of PERT (Program Evaluation and Review Technique) in the 1950s to help manage the Fleet Ballistic Missile program, new management techniques and organization constructs have been created to meet the challenge. The Pentagon's attempt to undertake massive, horizontally integrated system-of-systems projects indicates that it is again time to review the tools and organizations at hand.

A series of workshops, cosponsored by CSIS's Defense-Industrial Initiatives Group and MIT's Security Studies Program with the support of the Defense Department's Systems and Software Engineering Directorate, was held to address the challenges of today's complex defense and net-centric systems. The workshops provided several important insights covered in the chapters of this book, but two deserve emphasis here. First, there is an important distinction between complicated projects and complex projects, and although the tools exist to execute complicated projects, it is entirely uncertain that they are valid for complex projects. Second, many approach the problem of developing complex weapons systems as a technological challenge, but it is first and foremost an organizational and governance challenge. Elements of a project that are often external to a complicated project, such as the politics of coordinating large numbers of constituencies, are well within the bounds of a complex project and are often the cause of its non-linear behavior. Therefore, choosing the right organizational and governance structure when undertaking a complex project is one of the most important determinants of success.

Our goal was to begin the investigation of these critical topics and help form a community of interest capable of addressing the challenges. We hope that this volume provides a useful set of first steps.

ACKNOWLEDGMENTS

The editors wish to thank the Office of the Secretary of Defense (OSD), Systems and Software Engineering Directorate, for generously supporting the series of workshops that led to the writing of this book. Many thanks also to the Security Studies Program at MIT, particularly to Harvey Sapolsky and Michael Schrage, for collaboration that made these workshops a success and helped to move the book project forward. The workshops would not have been possible without the help of then CSIS Defense-Industrial Group colleagues David Scruggs and Judy Siegal. CSIS research associate Greg Sanders deserves special mention for his outstanding research support. James Dunton and Donna Spitler at The CSIS Press provided invaluable feedback and played a crucial role in the book's production. Finally, thanks to all who participated in the workshops and in the subsequent discussions; your insights generated the intellectual capital that served as a basis for this book.

INTRODUCTION
FRAMING THE COMPLEXITY CHALLENGE

GUY BEN-ARI AND MATTHEW ZLATNIK

Complexity increasingly affects the organizations that produce, acquire, and use military systems. Government and the private sector must find ways to manage it.

America has not contended with a technologically superior military since World War II, a tribute to the country's ability to harness its industrial base to its general technological sophistication and to the relative capital intensity of its economy. Implicitly or explicitly, U.S. defense strategy rests on a continuing ability to develop, acquire, and properly deploy advanced military technology. Although it provides superior tactical and operational capability, technological innovation also entails increased complexity. To keep its technological edge, the United States must find ways to accommodate not only the pace of technological change, a consolidating industrial base, limited budgets, and more challenging operating environments, but also the complexity generated by the interaction of these elements.

Historically, the practical approach to handling the complex has been to deconstruct it to the more manageable "complicated." Complicated systems are characterized by their large scale and a multitude of moving parts or actors that are highly dynamic—that is, that constantly interact with and affect one another. As such, defense planners have developed the working assumptions that a complicated system is ultimately understandable through the decomposition of its constituent elements and that the consequences of actions taken within it—cause and effect—are predictable to a great extent.[1] Traditional defense planning has proven quite adequate for dealing with complicated challenges. Defense planners have developed techniques such as doctrine and training and modeling and simulation to analyze them and find potential solutions.

Like complicated environments, complex environments are characterized by large scale and a multitude of interrelated elements. However, even an in-depth familiarity with each of these elements does not impart an understanding of the system or the environment as a whole.[2] Furthermore, complex systems and environments involve changes in the state or behavior of their various elements that cannot be predicted in detail. As described by David Alberts and Richard Hayes, "Couplings exist between apparently distant (and disparate) elements of the operating environment and little or no coupling exists among elements thought by some to be closely coupled."[3] Furthermore, even the smallest changes may initiate large variations in the resulting pattern of behavior and thereby affect strategic outcomes. As a result, flexibility and resilience become key characteristics of managing complexity; getting everything right the first time becomes a pipe dream. Policymakers, planners, and managers must employ multiple approaches, and the actors involved must be ready to alter their behavior in order to cope with unintended consequences and novel strategies implemented.

Empirically, using the techniques for addressing complicated challenges to solve the complex challenges of the twenty-first century simply does not work. Defense programs, while incorporating the best efforts of thousands of professionals, routinely are delivered late and cost more than expected. Because of the desire to achieve the synergies that complex systems allow, disaggregating the apparent problem is not an option; in the famous words of ecologist Garrett Hardin: "We can never do merely one thing."[4] Therefore, tools are required that enable coping with unexpected difficulties and taking advantage of unforeseen opportunities. As Columbia University professor Robert Jervis has observed: "When any element [of a system] underperforms or one part of a policy fails, other parts must be added or improved; when unforeseen interactions appear, several alterations are likely to be necessary."[5]

Thus, while policymakers, planners, and managers need to try to anticipate the reactions to the actions they design and implement, they also must have the ability to respond effectively if their assumptions are proven to be incorrect and to continue to look out for future repercussions. "Problems are almost never solved once and for all," writes Jervis; "initial policies, no matter how well designed, rarely can be definitive; solutions will generate unexpected difficulties."[6] Dealing with complexity therefore requires not merely flexibility to deal with a specific problem, but continuous flexibility to deal with continuously evolving problems.

Aside from its impact on the ability to acquire and deploy equipment, complexity affects innovation itself—that is, what can in fact be developed and acquired. Innovation is stimulated by competition; the latter provides incentives to entrepreneurs and companies to invest in new ideas, based on the knowledge that these investments will generate an edge in a competitive marketplace.[7] Therefore, the consolidation in parts of the industrial base supporting defense might be expected to reduce competition and therefore innovation. Perhaps by way of responding to this, the Defense Department allocates almost two-thirds of its contracted research and development spending through competitive processes,[8] which in turn bring additional complexity to the development and acquisition process. Defense policy toward innovation and competition is also challenged by the high rate of technological innovation and the increasingly blurred demarcation between purely military and purely civilian technology. Clearly, the transition to "merely" complicated is not made so easily.

The private sector faces a similar ongoing challenge, in the form of managing a torrent of information, rapidly changing market conditions, shrinking periods of competitive advantage, fickle capital markets, and global customers, suppliers, competitors and workforces. A recent report by McKinsey & Company showed that although complexity increasingly is unavoidable for many companies, those that embrace it by building the right processes, skills, and culture to manage it are developing competitive advantages, including an ability to create more value, increased resilience, and greater uniqueness (that is, it becomes harder for others to imitate what the company is doing).[9]

One approach to addressing complexity that is typical to the private sector is to improve the ability of an organization to understand and respond to changes in its environment by becoming a "learning organization." Such organizations can learn and adapt to changes in the environment and ultimately bring about their own continuing transformation.[10] Responsibility and accountability are pushed downward in the hierarchy, with incentives and rewards realigned correspondingly. Information technology systems are installed to share knowledge about the internal and external environments, allow workers to communicate more easily, and help the organization adapt easily to change. Shorter product design and production cycles and smaller inventories allow quicker reaction to changes in demand. Rigorous financial analysis is undertaken to enable a close focus on shareholder value and identify high-return activities.

But rather than simply maximizing shareholder value, government must reconcile several sometimes-incompatible goals. As such, private-sector management approaches may be insufficient. Defense policymakers must create capabilities, accomplish missions, and meet commitments within economic, legal, and political constraints, both domestic and international. Defense planners face difficult cost-benefit decisions, especially with long-lead-time, rarely used items: one more DDG-1000 destroyer, 10 more F-22 fighter aircraft, or another infantry battalion? In its final report from January 2006, the Defense Acquisition Performance Assessment (DAPA) project stated: "The Acquisition System must deal with external instability, a changing security environment and challenging national security issues. The Department must be agile—to an unprecedented degree—to respond quickly to urgent operational needs from across the entire spectrum of potential conflicts."[11]

Paradoxically, buying the most technologically advanced—and therefore increasingly complex—equipment may widen the military technology gap with potential adversaries but may not provide significant advantages in ongoing operations. Current and future adversaries are no fools and will find ways to engage the United States that negate the advantage provided by high technology. Policymakers must therefore be concerned not only with how to handle the development, acquisition, operation, and maintenance of increasingly sophisticated technology, but also with how an enemy might seek to defeat state-of-the-art equipment asymmetrically. Training and leadership cannot be neglected in the effort to harness technological wonders. The element of complexity is as critical in today's military training and leadership development as it is in technology development.

The existence of complexity as a continuous component in current and future defense environments means that those affected by it face two basic options. The first is to view complexity as an obstacle that creates undesirable effects such as cost increases and schedule slippages without conveying sufficient advantages; as such, attempts must be made to remove or overcome complexity through simplification, deconstruction, downsizing, or streamlining. The second opinion is to view complexity as an opportunity to be managed, possibly even exploited.

To examine the pros and cons of each approach in the defense environment and to initiate a process that will help frame the challenge of complexity for the Department of Defense and identify key intellectual building blocks for thinking about managing that complexity, the Center for Strategic and International Studies' Defense-Industrial Initiatives Group and MIT's Security Studies Program hosted a series of workshops. Under the overall title, "Organizing for a Complex World: Designing, Developing and Deploying Complex Weapon and Net-Centric Systems," five workshops were held covering the following issues:

- What is a "smart buyer" in a complex world?
- Defining complex systems: eccentric or net-centric?
- Competition and innovation in a complex environment
- Measuring complexity
- Applying new thinking about complexity

This book presents the key intellectual capital collected during these workshops. It aims to provide clear and detailed views on complexity in a national defense context and present several examples of specific areas where new thinking is required to address complex challenges. Any research on complexity must be careful to avoid two common pitfalls: a lack of clarity and a strained imposition of one school of complexity thinking in the examples and case studies presented. The central task of the workshops—and therefore of this book—was to elucidate what complexity is and to analyze different elements of it within the context of national defense. Therefore, the chapters in this book were chosen to represent the topics addressed at the various workshops and in subsequent discussions on the key elements of complexity. They were written by a mix of government, academia, and private sector experts who have regularly grappled, and continue to grapple, with the challenges posed by complexity.

Michael Schrage highlights the underappreciated importance of "complex systems" governance. Great engineering successes, he notes, result as much from "good governance" as successful project management and technical prowess. His chapter looks at best practices—and worst practices—in the commercial sector for governance design of national security procurements. Complex systems governance should focus on how to understand and manage risk, and only through the evolution of governance structures can better development and procurement outcomes be achieved. Please note that this chapter was written before the fall of 2008 and what now appears to have been massive failures of financial institution governance, oversight, and management of complex derivatives risk.

Harvey M. Sapolsky describes how military acquisition has evolved from the government-owned arsenal to the recent Lead Systems Integrator model. He notes the gradual shift from public to private execution not only of manufacturing but also of program management, technical direction, and requirements definition, resulting from changes in the political, economic, and technical realms. In an increasingly complex environment, manufacturing and design expertise and the ability to analyze requirements have shifted to the private sector, perhaps in an attempt to align responsibility and accountability.

Jeffrey A. Drezner describes how complexity affects innovation and competition. Competition is a necessary but not sufficient condition for innovation, which in turn results from firms' attempts to remain viable in a competitive marketplace with buyers who have a choice of suppliers. As technological, organizational, and environmental complexity has grown, firms have responded by consolidating, which may reduce competition and therefore innovation, at least at the prime contractor level. However, the need for innovation will continue as the environment creates new warfighting needs.

To round out the exploration of governance and complexity, Eugene Gholz assesses how well different types of organizations provide the attributes needed for successful systems integration. Each organization type—government laboratories, federally funded research and development centers, and private corporations—excels in some attributes but falls short in others. Policymakers therefore must decide which aspects of a development effort are most important and select a governance organization based on the model that best delivers those aspects.

Within an organization, Jeremy M. Kaplan characterizes the challenge as how to coordinate multiple, simultaneous development efforts, so that each resulting system functions as intended, is interoperable, and contributes effectively to an overall system-of-systems optimized for a variety of missions. A network-based organizational structure best provides effective project management, balancing the costs and advantages of hierarchy and decentralization. Within the organization, each system-of-systems authority—responsible for the outcome—needs to be supported by a system-of-systems engineer and an environment that enables subsystem developers to coordinate among themselves and make appropriate cost-benefit analyses. A networked, collaborative environment and a network-centric culture are required to make this organizational model work.

Douglas O. Norman describes how turning operational requirements into an executable engineering program may degrade the requirements such that something is lost and the result is suboptimal. A classical response is to define requirements ever more carefully to make sure that what is delivered is precisely what is expected, which adds expense and time, perhaps resulting in something that is obsolete when delivered. Responses to this problem, such as developing in isolation or indulging in negotiations, further make it difficult to deliver operational value, by causing environmental developments to be ignored or by adding time and administrative complexity to the development process. A user-focused marketplace might return emphasis to delivering value.

Marco Iansiti argues that the Lead Systems Integrator concept, as applied to the Army's Future Combat Systems (FCS) project, includes many of the managerial techniques, competencies, and strategies used in successful megaprojects. Such projects are characterized by technological and organizational complexity, uncertainty of the environment, and external factors affecting outcomes, often factors out of direct control of the project managers, and a large number of distributed stakeholders. With such attributes, aggressive, adaptive management strategies are required to reduce the risk of failure. A number of successful strategies were identified through a study of 31 technology-intensive megaprojects, and many of those have been applied to the FCS through the Lead Systems Integrator concept.

W. Scott Gould and Julie M. Anderson describe how the private sector can work with the government acquisition system, which is best thought of as a network of public and private stakeholders. To better understand government decisions, network participants must share information and combine forces. To succeed, vendors must demonstrate that they understand not only the contract requirements but also the environment, the goals of the program, and the government's desired outcome. Similarly, the public sector must offer more flexible, outcome-based contracts, provide guidance rather than control, and apply private-sector change management techniques.

David H. Dombkins reminds us that management of complex projects requires attributes beyond technical skill, including leadership qualities resulting from a combination of professional and personal experiences and personality traits. Such characteristics take years to cultivate, and only a limited number of project managers will ever be able to handle the most-complex projects. High-potential candidates should be identified early and put on a development fast track.

With ever more sophisticated systems, organizations, and military missions, complexity has become a fact of life for the Department of Defense and the industrial base that supports it. To provide continuing innovation, these entities must concern themselves with how increased complexity—of systems, the operating environment, the acquisition organization, etc.—affects innovation and how to handle it. The defense capabilities—whether goods, services, or processes—required for current and future operations must be designed, built, and procured in a manner that enables their integration across the battlefield. Complexity increases the difficulty of this task, and new organizations and management approaches will be needed to address it.

NOTES

1. David S. Alberts and Richard E. Hayes, *Planning: Complex Endeavors* (Washington, D.C.: Department of Defense Command and Control Research Program, 2007), pp. 11–12.

2. Elizabeth Garnsey and James McGlade, *Complexity and Co-Evolution: Continuity and Change in Socio-Economic Systems* (Northampton, Mass.: Edward Elgar, 2006), p. 3.

3. Alberts and Hayes, *Planning: Complex Endeavors,* p. 16.

4. Garrett Hardin, "The Cybernetics of Competition: A Biologist's View of Society," *Perspectives in Biology and Medicine* 7 (Autumn 1963): 80.

5. Robert Jervis, *System Effects: Complexity in Political and Social Life* (Princeton, N.J.: Princeton University Press, 1997), p. 294.

6. Ibid.

7. See, for example, Joseph Schumpeter, *The Theory of Economic Development* (Cambridge: Harvard University Press, 1934); Robert Gordon, "The

United States," in *Technological Innovation and Economic Performance,* ed. Benn Steil, David G. Victor, and Richard R. Nelson (Princeton, N.J.: Princeton University Press, 2002).

8. CSIS analysis of Federal Procurement Data System (FPDS) data, 2007.

9. Suzanne Heywood, Jessica Spungin, and John Turnbull, "Cracking the Complexity Code," *The McKinsey Quarterly,* May 2007.

10. Donald Schön, *Beyond the Stable State: Public and Private Learning in a Changing Society* (Harmondsworth, England: Penguin, 1973), p. 28.

11. Department of Defense, *Defense Acquisition Performance Assessment Report* (2006), p. 7, http://www.acq.osd.mil/dapaproject/documents/DAPA-Report-web/DAPA-Report-web-feb21.pdf.

MAKING GOVERNANCE MATTER MORE
OVERSIGHT, INSIGHT, AND FORESIGHT IN COMPLEX SYSTEMS PROCUREMENTS

MICHAEL SCHRAGE

Give them the third best to go on with; the second best comes too late; the best never comes.
  —Robert Watson-Watt, *Three Steps to Victory*[1]

Championing "third-best" solutions in wartime might appear grossly cynical or naive. Yet this innovation heuristic faithfully reflected Robert Watson-Watt's design sensibility for complex systems development. The Royal Air Force's "third-best" approach to radar both as emerging technology and as essential systems ingredient became one of the great stories—and great clichés—of successful military innovation.

But even great clichés should not obscure larger truths. For all Watson-Watt's brilliance, the institutional reality was Air Marshal Hugh Dowding, who—in concert and coordination with the Air Ministry's Committee for the Scientific Study of Air Defence—oversaw radar's funding, experimentation, and implementation as a complex network technology. That integrated system—from the Chain Home towers to the Bentley Priory filter rooms to the Spitfires—completely transformed the Royal Air Force's Fighter Command and Great Britain's security.

Where Watson-Watt was the paradigm of an intrapreneurial innovator who charismatically manipulates bureaucracies and technologies to achieve his ends, "Stuffy" Dowding was a near-caricature of Old Boy institutional oversight and accountability. Dowding, along with key colleagues such as Henry Tizard, effectively determined the supervisory structures and testing regimes that enabled Watson-Watt's "third-best" innovations to consistently succeed. Superior governance—not just superior technical ingenuity—made radar a war winner.[2]

Governance is the focus of this chapter. Wherever complex military systems have been successfully procured and deployed—Sidewinder, Polaris, the U-2, Stealth technologies, intercontinental ballistic missiles (ICBMs), AC-130 gunships, Predator unmanned aerial vehicles (UAVs), global positioning system (GPS)—the narrative of success has typically centered on the outstanding quality of technical leadership. The Bill MacLeans, Kelly Johnsons, Bernard Schrievers, Brad Parkinsons, and William Raborns are the innovation heroes—the Watson-Watts of their domains.[3]

Although these narratives are technically accurate, they disproportionately—and misleadingly—downplay the role of "governance" in systems success. Again, the institutional reality is that Kelly Johnson's U-2 depended upon shared understandings around oversight reached with the Central Intelligence Agency's Richard Bissell;[4] Col. Bradford Parkinson's global positioning system satellite infrastructure required innovative governance mechanisms improvised and imposed by Defense undersecretary Malcolm Currie;[5] and before PERT (the program evaluation review technique) was even a gleam in Admiral William Raborn's eye, the most important step in launching Polaris was Admiral Arleigh Burke's establishment of the Special Projects Office. Burke literally created a new governance structure to have his navy build its ballistic missile.[6]

The point is not to diminish the importance of competence, courage, and charisma in leading complex systems innovation. But any dispassionate review of complex systems histories in the postwar era suggests that governance "best practice"—however defined—is perniciously underappreciated as a critical success factor.

More dangerously, the purpose and power of governance in systems acquisition has been obscured in favor of celebrating technical leadership. The resulting imbalance undermines the value of governance, technical leadership, and the systems themselves. This typically leads to mismanaged expectations, poor investment, and unhappy outcomes.

By consistently emphasizing functional oversight over strategic insight and foresight, for example, governance in defense acquisition and procurement has been pathologically rebranded as either overhead or adversarial to technical leadership. That is counterproductive and unsustainable. The future of complex systems development requires national security policymakers to revisit and revise their governance principles and practices.[7]

The challenge is not "organizational reinvention" around governance, but rethinking how complex systems leadership should identify and manage complexity risk. Effective governance disciplines this rethinking. Risk management—as opposed to ensuring "conformance to requirements"—becomes an essential operating principle for aligning governance with leadership.

Make no mistake: Leadership is not governance; governance is not leadership. Their value asymmetry is inherent. Brilliant systems development teams can—and frequently do—succeed in spite of perfunctory, bureaucratic, or even obstructionist oversight. Conversely, even the most energetic and insightful governance cannot drag incompetent technical leadership across a successful systems finish line. Then again, energetic and insightful governance would never allow mediocre—let alone incompetent—technical leadership to command a complex system development.

Governance should not be an operational substitute for leadership any more than leadership can substitute effectively for management. The spans of control fundamentally differ. A Tizard cannot substitute for a Watson-Watt. An Arleigh Burke should not redraw Raborn's Polaris PERT charts. Yet, historically, governance has individually and institutionally influenced—for better and worse—leadership's ability to successfully deliver complex military systems.

The important questions consequently become the following: How can governance and leadership best align to successfully manage the inherent and emergent risks of complex systems development? What should or must accountability mean to make that alignment effective, productive, and successful? As systems complexity evolves, how must or should that alignment evolve?

The best answers will not be found in the rapidly expanding literature on complex systems management. The most useful insights and information will instead emerge from the even-faster-growing literature on corporate governance. Corporate governance provides the richest reservoir of competitive ideas about alignment. How corporate boards and independent directors should best oversee management performance and mitigate enterprise risk provokes intensifying arguments about appropriate governance.[8]

The public crisis in corporate governance—from Enron to WorldCom to Tyco to Citigroup—invited radical reform. Yet litigation, regulation, and legislation ostensibly intended to improve governance have failed to evoke meaningful consensus around the board's appropriate role. On the contrary, they have further polarized discussion around transparency, accountability, fiduciary duty, and independent judgment to ensure both legal compliance and marketplace demands. They have sharpened conversation about desired quality and character of nonexecutive directors. Governance has become a global battleground where firms, investors, and policymakers fight over the rules of corporate control. It is a battlespace worthy of E-Ring ISR (intelligence, surveillance, and reconnaissance).

The new institutional reality is that these fundamental disagreements increasingly shape—for better and worse—how publicly traded enterprises structure and oversee complex systems innovations. A Watson-Watt or Bissell would immediately grasp the risk dynamics those disagreements inject into internal systems development debates.

Of course, governance of complex military systems procurement is not the governance of publicly traded firms. The market tests and legal obligations are different. Corporate directors are fiduciaries; Congress, inspectors general, and assorted defense bureaucracies are not. Nonexecutive corporate directors must be free of conflicts of interest; congressional representatives and defense officials are often riven with them. Boards are empowered to hire, fire, and compensate executives; no comparable flexibility exists in national security systems governance. Directors supposedly represent the firm's shareholders; the interests defense overseers represent are ordinarily less than clear.

Nevertheless, the similarities between corporate and defense systems governance overwhelm the differences. Virtually every significant governance challenge confronting defense procurement has distinct counterparts and analogs in corporate governance. Practically every substantive debate about the corporate governance future—principle vs. rule-based oversight, independence, transparency, executive succession, the role of external advisers, compliance, accountability, enterprise risk, executive compensation, informed business judgment—is directly relevant to governance challenges in defense.

This is not the conventional call for importing better "business practice" into complex defense procurements. That would be near impossible. No consensus "best practice" for corporate governance exists. Whether corporate boards should have "nonexecutive chairmen" or use decision support technologies for boardroom deliberations, for example, remains in hot dispute among elite directors and leading governance scholars.[9]

But in the turbulent wake of the Sarbanes-Oxley legislation of 2002 and the plethora of regulatory and litigatory rulings since, corporate boards have become living laboratories for governance experimentation. Boardrooms are now prototypes and beta sites for benchmarking governance reform impacts on complexity management. This presents an enormous learning and risk management opportunity for defense policymakers and practitioners. Real-world governance insights are ripe for the taking.

For example, Warren Buffett, arguably the world's most successful long-term investor, maintains that a board's most important responsibility is removing chief executive officers who are simply not good enough. What makes a board's job particularly difficult, he has noted in workshops on corporate governance, is what to do with CEOs who rate a 6 or 7 on a 10-point scale. Are the replacement risks worth it?

For many complex systems development efforts, this kind of governance principle might be appropriate. Quickly terminating poor leadership could be a welcome good-governance measure of effectiveness.

By contrast, governance guru Ira Millstein, a lawyer who has advised boards at the world's largest companies, asserts that productive partnership with chief executives is paramount. Executive accountability is important, he agrees, but directors should be willing collaborators to help their CEOs succeed.[10] How could—how should—more collaborative governance enhance oversight and evaluation of complex systems development? How might that be measured?

Whether defined by explicit rules or determined by general principles, what is quickly clear is that effective governance requires cultivation and management of meaningful expectations. What should "good governance" or "world-class" governance look and feel like? What governance "brand" would likely work best for the program?

Unclear and uncertain expectations dominate how boards seek to simultaneously impose executive accountability while overseeing enterprise risk. Remarkably few boards clearly articulate—let alone actively implement—expectations-setting policies. But the need for mission clarity grows especially important for organizations confronting increased complexity. Effective *governance* of complex systems development initiatives demand different sensibilities and expertise than are required for *leading* them.

This is where complexity's impact and importance are fundamentally misunderstood. Yes, technical leadership must be able to identify, manage, and tame complexity. Yes, technical leadership should also be able to identify, manage, and tame the complications. But distinctions between "complex" and "complicated" often prove to be purely matters of semantics and perspective. A developer's "complex" system may be the user's "complicated" system—or vice versa.

Enhancing governance, however, renders these definitions operationally irrelevant. What governance must rigorously oversee—and be held to account for—are the levels of risk the organization confronts. The greater the perceived or actual risk, the greater governance's curiosity and concern must be. How much is enough? How much is too much? Why?

If increased upstream complexity measurably increases downstream risk, then governance must evaluate whether the magnitude of either increase is worth it. If additional complexity creates greater systems utility with only marginal additional risk, then governance might not care. Of course, added complexity may even reduce overall risk; that is the goal of diversification.

Overall risk—not overall complexity—is the metric that should matter most to governance. Risk management—not systems complexity—most clearly delineates the respective roles and obligations of technical leadership and programmatic governance. Technical leadership designs and determines the complexity "trade space." Governance must be comfortable—if not confident—that risks created by complexity are foreseeable, manageable, and acceptable. Where leadership "defines" the technical vocabularies of complexity, governance "owns" the organizational definitions of risk. Failure to usefully define and align these vocabularies creates risks all their own.

The great contemporary example of governance failure confusing complexity with risk is the subprime mortgage/collateralized debt obligation (CDO) global financial crisis. By July 2008, the so-called mortgage meltdown had cost financial institutions worldwide almost half-a-trillion dollars in write-downs and loan losses. Financial institutions with rigorous risk-management controls and effective corporate governance sustained some damage, but several of the world's biggest banks—with "elite" and "sophisticated" corporate boards—collectively lost well more than $175 billion.[11] Even by Pentagon standards, those sums are significant.

A detailed recounting of this ongoing debacle lies beyond this chapter's scope. But the underlying complexity challenges mortally wounding so many leading financial institutions appear remarkably similar to those confronting the Office of the Secretary of Defense, the military services, and large-scale integrators. Synthetic derivatives and fire-control software are products of comparable complex systems engineering design.

Wall Street "quants" (financial engineers) are nicknamed "rocket scientists" for good reason. Tools, technologies, and techniques used to iteratively test complex systems innovation have a lot in common. They are typically computationally intensive and networked. They require expensive investments in technical talent and infrastructure. Success is predicated on cost-effective complexity management married to a disciplined willingness to identify and mitigate emergent risk.

Yet despite great skill, expertise, and monies invested in complex financial innovation, top-tier institutional innovators abjectly failed to assess real risks. Real-world risk exposures proved orders of magnitude larger than internally discussed. To be fair, misjudging financial risk happens all the time. Mistakes and miscalculations—even large ones—occur with alarming frequency. The best risk management systems are not infallible.[12]

But the failures here went well beyond finance. The institutional reality was that governance—corporate boards at Citigroup, UBS, Merrill Lynch, and Bear Stearns, among others—had completely and disastrously failed to detect the enormously increased risks that had been acquired with increased innovation complexity. These failures publicly exposed the inability of boards to effectively oversee complexity risk. They precipitated a crisis in the global financial system.

This brief critique highlights basic confusion surrounding real-world expectations of governance. Can boards adequately represent shareholder interests if they are demonstrably ignorant or unaware of complexity-created risk? Just how well can complex military systems governance reflect taxpayer and warfighter concerns if complexity risk is not assessed?

Ultimately, the questions resurface fundamental arguments about what defines effective governance. If governance cannot substantively help leadership manage and mitigate risk, then what is it good for? Maybe governance should downplay its risk assessment competence. As systems complexity increases, should governance become more involved or less intrusive in assessing enterprise exposure? Continuing quantitative and qualitative changes in complexity force ongoing reevaluations of governance's role and value in risk oversight.

Once again, painful experiences from finance offer actionable insight. UBS, a Swiss bank globally aggressive in the CDO innovation marketplace, suffered losses approaching $40 billion on its complex innovations positions and portfolio. Its corporate board consequently commissioned a report to identify what went wrong. In April 2008, the bank published a comprehensive forensic analysis describing its litany of loss.[13]

The findings were damning. The report declared senior UBS management had a "failure to demand holistic risk assessment"; a "failure to manage [its] agenda"; and a "lack of succession planning."[14] Additionally, the report excoriated the firm's risk management controls and testing methodologies, asserting "complex and incomplete risk reporting," "lack of substantive assessment," "inadequate systems," "lack of strategic coordination," and an "inability to accurately assess valuation risk on a timely basis."[15]

The report also detailed perverse incentives—including "asymmetric risk/reward compensation" and "insufficient incentives to protect the UBS franchise long term"[16]—that exacerbated the bank's other institutional flaws. Goldman Sachs's chief executive and the New York Federal Reserve Bank president both independently observed that the UBS report sparked reexamination of risk management practices and cultures in financial service firms worldwide.

Ironically and perhaps perversely, the report declared UBS's governance framework "appropriate and with clear allocation of responsibilities." The review insisted that "the overriding issues relate principally to implementation and effectiveness"[17]—which, of course, begs the question of what role governance should have played to ensure effective implementation.

Whether informed readers agree with those judgments or not, the institutional reality is that corporate governance commissioned and coordinated this external assessment of internal failures. Moreover, these assessments invited credible criticism of the board's own performance as an overseer of executive leadership, complex systems management, and enterprise risk. Four UBS directors stepped down shortly after the report was issued.

Indeed, the report's *ex post* nature recalls the health care jibe that pathologists have by far the greatest knowledge of medicine but, unfortunately, too late. UBS investors would likely have preferred a board that had proactively identified the vulnerabilities and taken steps to fix them. If more prudent governance had reduced exposures by only 15 percent, bank losses would have been cut by more than $5 billion. Even marginally better governance can be worth billions.

Several expert observers of complex financial securities markets argue that innovative instruments require more interactive governance oversight. Rick Bookstaber, a pioneering Wall Street quant and former Morgan Stanley and Salomon Brothers risk management officer, has suggested that boards explicitly direct management to run "stress test" simulations and scenarios to challenge the firm's risk assumptions.[18] Instead of reviewing strategies, directors—drawing upon their breadth and depth of global experience—should test how resiliently complex innovations might respond to their scenarios.

Such recommendations typically draw accusations of "micromanagement" and board-level "interference" in operations by governance critics who believe that nonexecutive directors should not be "backseat drivers" of the firm. But drawing bright-line distinctions between "advice," "engagement," and outright "challenges" in how governance reviews risk-related management proposals is inherently awkward. Governance and leadership may reasonably have radically different risk assessments because of their different perspectives. Helping create shared awareness of risk dimensions is arguably a rational oversight criterion.

Requiring that leadership's risk management models endure interrogative testing—as opposed to passive auditing—seems less board-level "micromanagement" than fulfilling governance's duty to improve management's risk awareness. By definition, innovations around complexity inject nontraditional risks for identification and consideration, which suggests that complex innovation systems should invite greater governance scrutiny, not less.[19]

Encouraging novel testing regimes to anticipate and ameliorate novel systems risk has, in fact, a proven past in successful governance of military systems development. Before World War II, Henry Tizard and his colleagues successfully pushed the Royal Air Force into a series of exercises to improve ground control of fighter operations. Experiments linking radio to radar became the nucleus of "ground control intercept" capabilities; sparked the discovery of "the Tizard angle" to rapidly calculate fighter flight paths; and launched the "filter rooms" that became the integrated command-and-control hubs for Fighter Command.[20]

Without diminishing in any way the technical achievements of Watson-Watt, Rowe, and the Air Ministry's Telecommunications Research Establishment, the insight and foresight to experiment and test radar more as a "system" than a "technology" proved a triumph of governance. Forty years later, the Pentagon's insistence—over service resistance—that "pseudolites" and other innovative testing schemes be supported to make GPS a more flexible and resilient platform proved prescient.[21]

Comparable stories can be found in ICBM and Polaris developments. As schedules and operational requirements shifted, "conforming to requirement" became less important relative to the ability to adaptively respond to—and learn from—challenging tests. Governance was less about a traditional auditing and oversight function than a challenge to leadership to demonstrate measurable progress through test.

As the historical record indicates, technical leadership of successful complex systems procurements were guided by governance that believed the military did not get the systems that were originally "designed"—instead, they got the systems that were effectively tested. Successful governance oversaw the design and development of successful testing.

It is sobering to acknowledge that the 68-page May 2008 *Report of the Defense Science Board Task Force on Developmental Test and Evaluation* does not once mention the word "governance."[22] Decoupling "developmental test and evaluation" from "governance" denigrates the history of complex systems development. More dangerously, it undermines conversations about the roles governance can play in balancing leadership accountability and risk management.

Focusing on fundamental improvements in the procurement and acquisitions process is undeniably important. But it beggars belief that fundamental improvements of process can be achieved—let alone organizationally sustained—without comparable focus on fundamental improvements in governance. From the Packard Commission on, there has been no shortage of proposed procurement reforms. The lingering question remains whether procurement reform inherently requires governance reform. It is noteworthy that the "governance" reforms defining the Goldwater-Nichols reorganization of command seemingly fade into insignificance in the context of procuring complex military systems.

To the extent that governance should become a more important mechanism for managing complex systems risk, the defense community needs to have almost exactly the same arguments and debate that now dominate corporate governance. Effective governance in the face of increased complexity increasingly requires three kinds of sight:

- **Oversight.** Who—and how—should governance hold leadership accountable for performance? What tests can we propose to best achieve this?

- **Insight.** What does governance learn and share about the nature of the system's complexity that improves leadership's chances to succeed and the procurement's chances to be successfully deployed? What experiments can we propose to best test this?

- **Foresight.** Given existing and emerging complexities, what are probable and possible impacts—for better or for worse—on the larger national security eco-system? What simulations and scenarios can we run to explore this?

Any serious discussion about determining optimal blends of insight, foresight, and oversight for governance should improve the process, performance, and productivity of defense acquisition. Of course, there is no avoiding the *quis custodiet ipses* issue: how does governance determine its own value, effectiveness, and utility?

That question evokes two governance anecdotes. The first comes from the late John R. Pierce, a prominent Bell Labs engineer who oversaw the successful and pioneering launch of Echo, America's first telecommunications satellite in 1960: "New projects, once fueled and on the road, suffer far more from backseat driving than from neglect," he observed. "A chief reason Echo succeeded was that the wrong people did not realize it was important until it was too late for them to interfere with it."[23]

By contrast, General Leslie Groves—whose complex systems Manhattan Project had the good fortune to be governed by men such as James Conant and Vannevar Bush—publicly commented:

> A major key to our success was this simple top organization and the quality of men who formed it. Meetings were always held at [Groves's] office . . . chaired by Bush . . . . But as time went by and I became much more familiar with our operations than any of the others, it became more and more a question of approval and discussion rather than of decision.[24]

Achieving a healthy balance between "Hippocratic" governance—governance that, first, does no harm—and governance where performance-driven trust is successfully attained should be an institutional aspiration for defense. Echo, the atom bomb, and radar were indeed the complex systems of their day.

Watson-Watt's admonition about technological innovation applies with equal import to governance innovation: the second best comes too late and the best never comes. Let us go with the third best.

NOTES

1. Robert Alexander Watson-Watt on the "cult of the imperfect," in *Three Steps to Victory* (London: Odhams Press Ltd., 1958), p. 74.

2. Watson-Watt, *Three Steps to Victory*; Ronald Clark, *Tizard* (Cambridge: MIT Press, 1965); David Zimmerman, *Britain's Shield: Radar and the Defeat of the Luftwaffe* (Stroud, Gloucestershire: Sutton Press, 2001); Louis Brown, *A Radar History of World War II: Technical and Military Imperatives* (New York: Taylor & Francis, 1999).

3. Arguably, America's most innovative organizational approach to the role of governance in complex systems research, development, and procurement was the World War II creation of the Office of Scientific Research and Development (OSRD) under Vannevar Bush in the summer of 1941. The OSRD operationally superceded the advisory and funding role of the National Defense Research Committee, which had been set up by Bush in 1940. Irvin Stewart, *Organizing Scientific Research for War: The Administrative History of the Office of Scientific Research and Development* (Boston: Little, Brown, 1948).

4. Clarence L. Johnson, *Kelly: More Than My Share of It All* (Washington, D.C.: Smithsonian Institution Press, 1985); Richard Bissell, *Reflections of a Cold Warrior: From Yalta to the Bay of Pigs* (New Haven, Conn.: Yale University Press, 1996).

5. Personal Communications, 2005; "Brad Parkinson, Electrical Engineer," an oral history conducted in 1999 by Michael Geselowitz, IEEE History

Center, Rutgers University, New Brunswick, N.J., http://www.ieee.org/portal/cms_docs_iportals/iportals/aboutus/history_center/oral_history/pdfs/Parkinson379.pdf.

6. Harvey Sapolsky, *The Polaris System Development: Bureaucratic and Programmatic Success in Government* (Cambridge: Harvard University Press, 1972).

7. Although governance issues are raised, for example, in the Department of Defense's *Defense Acquisition Transformation Report to Congress* (February 2007, http://www.defenselink.mil/pubs/pdfs/804Reportfeb2007.pdf), neither the definitions and approaches to "governance" nor the absence of discussion of challenges posed by complex systems address the central concerns raised by this chapter.

8. See, for example, Ira Millstein and Paul W. MacAvoy, *The Recurrent Crisis in Corporate Governance* (paperback) (Stanford, Calif.: Stanford Business Books, 2004); Robert A.G. Monks and Nell Minow, *Corporate Governance* (Chichester, England: John Wiley & Sons, 2008); John C. Coffee, *Gatekeepers: The Professions and Corporate Governance* (New York: Oxford University Press, 2006); Roy Smith, *Governing the Modern Corporation: Capital Markets, Corporate Control, and Economic Performance* (New York: Oxford University Press, 2006); Stephen Bainbridge, *The New Corporate Governance in Theory and Practice* (New York: Oxford University Press, 2008).

9. Sir Adrian Cadbury, *Corporate Governance and Chairmanship: A Personal View* (New York: Oxford University Press, 2002); Millstein Center for Corporate Governance Conference, June 2008.

10. "Ira M. Millstein: The Thought Leader Interview," *strategy+business,* Spring 2005, http://www.strategy-business.com/press/16635507/05109.

11. More specifically, Citigroup's declared losses exceeded $54 billion; Merrill Lynch's losses topped $51 billion; Morgan Stanley's approached $15 billion. The write-downs of Goldman Sachs, by contrast, were less than $4 billion. See John Gapper, "The Cost of a Wrong Turn," *Financial Times,* August 5, 2008.

12. See Rob Davies, "Reporting for Duty," *Risk* magazine, July 2008, http://www.risk.net/public/showPage.html?page=802925.

13. UBS, *Shareholder Report on UBS's Write-downs,* April 2008, from http://www.ubs.com/1/g/investors/shareholderreport.html.

14. Ibid., p. 35.
15. Ibid., pp. 39–41.
16. Ibid., p. 42.
17. Ibid., pp. 34–35.

18. Richard Bookstaber, *A Demon of Our Own Design: Markets, Hedge Funds, and the Perils of Financial Innovation* (Hoboken, N.J.: John Wiley & Sons, 2007); also personal communications.

19. See Riccardo Rebonato, *Plight of the Fortune Tellers: Why We Need to Manage Financial Risk Differently* (Princeton, N.J.: Princeton University Press, 2007).

20. Clark, *Tizard*.
21. Brad Parkinson interview conducted by Geselowitz.
22. Department of Defense, http://www.acq.osd.mil/dsb/reports/2008-05-DTE.pdf.
23. Fremont Kast and James Rosenzweig, eds., *Science, Technology, and Management*, Proceedings of the National Advanced Technology Management Conference, Seattle, Washington, September 1962 (New York: McGraw-Hill, 1963), p. 76.
24. Ibid., p. 33.

MODELS FOR GOVERNING LARGE SYSTEMS PROJECTS

HARVEY M. SAPOLSKY

Weapon acquisitions of even moderate scale and complexity require the creation of a community of interest among technologists, military users, and service managers. They have to believe in the feasibility of and need for their system enough to persuade the many others outside the project to assist in its realization. Among the others whose cooperation is needed are the Congress, other parts of the Defense Department including components of the sponsoring service as well as the Office of the Secretary of Defense, technical experts in and out of government, and the attentive media. Not even the blackest of the black programs in our decentralized governmental system can be totally free from need to gain the acquiescence of some who can delay or block its progress.

Skillfully expressed, political will alone can initiate weapon projects for the nuttiest of technical concepts, but eventually the project's technological failings will become known and alternative systems will be sought. To be successful, projects need to have political support and visible achievements—the required budgetary access and the independent organizational base that come with strong political backing as well as the expectation of a viable system that meeting a string of technical objectives engenders. Political backing buys time and resources while milestone achievements buy an opportunity to make a lasting contribution to national security. However, you usually do not get the latter without the former.

Those who have the responsibility for setting up projects need to think about both the project's political support and its associated technological capabilities. There are often difficult trade-offs to be made. The very things that might help a project gain political support may reduce its technical capacity. Some organizations may have better political skills and fewer technical skills than alternatives. The teams more likely to meet technological goals may not be able to attract the public support that would bring in funds necessary to make such advances. Both skills are vital, but do not come equally distributed.

Individual organizations have unique mixes of the necessary skills. It is impossible to prescribe which ones should be sought for a specific project without knowing them as unique entities. For planning purposes we should think in terms of sectors and their likely contribution to ensuring project success. The characteristics of sectors in terms of valued skills do change as society changes, but likely more slowly and less dramatically than would individual organizations where the presence or absence of particular persons could make a significant difference. When Admiral Hyman Rickover finally retired, the Nuclear Navy's political clout within the Navy and within government was substantially diminished. But the end of the Cold War, crushing though it was to the Nuclear Navy, impacted only moderately the military's overall prospects relative to other societal institutions.

The big choice is between the public and the private sectors—between government and industry—in being assigned responsibility for the tasks necessary to build weapons. There are four main tasks: setting *program requirements,* which involves determining the military goals for the system; providing *program technical direction,* which involves making the key technical trade-offs; controlling *program management,* which involves setting schedules, monitoring budgets, and selecting personnel; and *program technical execution,* which involves developing and producing the system. The result when these tasks are assigned alternative to government agencies or firms as they have been over the years are five basic governing models: *Arsenal, Contract, Weapon System Manager, Outsourcing,* and *Lead Systems Integrator.* The models discussed here are outlined in table 3.1.

ARSENAL

The arsenal model refers to the traditional way in which weapons were acquired. In the Age of Sail and long afterward, the Navy could not

Table 3.1. Program Responsibility Format Types

	Arsenal	Contract	Weapon System Manager	Outsourcing to Private Arsenal	Lead Systems Integrator
Program requirements	Government	Government	Government	Government	Industry
Technical direction	Government	Government	Government	Industry	Industry
Program management	Government	Government	Industry	Industry	Industry
Technical execution	Government	Industry	Industry	Industry	Industry
External environment	▪ Little commercial application of military tech	▪ Some commercial application of military tech ▪ Private sector pays better, can be more responsive	▪ Weapons become more complicated/complex ▪ Coordination of subsystems becomes important ▪ Large companies can leverage political support more	▪ Government begins to lose in-house tech capabilities ▪ Outsourcing becomes increasingly acceptable	▪ Loss of in-house government tech capabilities leads to inability to define what is possible

only specify the warship it needed, but could also accomplish the design, construction, and outfitting of the ship. The Navy made its own rope, forged its own anchors and cannons, and set its own technical standards. Later it developed the capability to design and produce its own aircraft. It was not totally self-sufficient, but was fairly close to it, as was the Army. The reason, of course, was that military needs are both specialized and episodic. Ships have to survive wars as well as the seas. Fortunately wars are infrequent, but because of that fact, the military's specialized technologies that they require have to be nurtured independently of the market. Firms cannot sustain themselves on the promise of purchases to occur in a future war. Thus the government sustains the vital technologies in its own facilities—arsenals and shipyards—between wars.

But for some emerging military-relevant technologies, the private sector proved to be the pacesetter. In aviation especially, entrepreneurs and inventors advanced the field not on the expectation of wars, but in the hope of commercial markets. Private equity built the industry that for more than half a century mainly served the military. The mobilization for the Cold War that followed so closely on the World War II mobilization persuaded many firms in that industry and others that the U.S. military in peacetime as well as in wartime was a big enough market in itself to sustain firms able to work on a broad spectrum of technologies.

CONTRACT

The contract model rather steadily came to replace the arsenal model for the acquisition of weapons during the Cold War. Scientists and engineers wanted to return to the unrestricted life of the private sector after their World War II service. The salaries were higher in industry and the bureaucracy less obtrusive. The armed services too favored industry contractors because they were more responsive to senior line officers than were the services' own arsenals and shipyards, which usually reported through a separate hierarchy and had close ties to congressional committees.

WEAPON SYSTEM MANAGER

As weapon complexity grew, so too did the realization that more systems thinking was needed in the weapon acquisition process. To function effectively, weapons require sensors, communications, platforms, crew training, spare parts, basing, and more. Beginning first in aircraft and missile development projects and in the Air Force, the structural shift was made toward employing large contractors as weapon system managers responsible for the administration and coordination of a network of contractors involved in the multiple subtasks required to create modern weapon systems. This shift helped simplify the government's role by passing much of the administration of the project over to the weapon system manager. It added the much-appreciated advantage of involving large, politically savvy contractors in a major way in the promotion of the system as well as in its development.

Over time the Department of Defense consciously shed much of its technical implementation and program management capabilities. Although not as intentional, it also began to lose its ability to provide program technical direction as these capabilities are interdependent. Technical direction skills are often acquired through involvement in bench work and/or lower-level management experience. The loss was accelerated by the end of the Cold War when there was pressure to reduce defense spending. The technical direction is almost exclusively a civil service activity while the setting of program requirements is primarily a military function. With reductions being imposed on defense, the military leadership wanted to preserve combat capabilities and sought the savings from the support side of the services, including much of the remaining program technical direction and program management staffs.

OUTSOURCING

The drive to the fourth model—outsourcing—came to fulfillment with a bipartisan consensus that private sector program implementation and management were to be preferred over government implementation and management for nearly every public activity. In defense, this consensus easily gained the tacit endorsement of the officer corps, whose collective opinion of defense civil servant dedication and competence did not have to fall much to match that of the average contractor. The collapse of the Soviet Empire seemed to confirm the belief of most American senior officials that markets were nearly always superior to government planning solutions. Few of these officials, however, took notice of the limitations inherent in defense markets, where the potential buyers are monopsonists or near monopsonists and where potential suppliers face significant barriers to entry because of classification restrictions and the specialized knowledge required to understand the buyer's needs and decisionmaking processes.

LEAD SYSTEMS INTEGRATOR

It was but a small step from the outsourcing framework to the Lead Systems Integrator framework. The requirement-setting task was controlled by military officers who were usually long on operational experience, but short on technical knowledge. With the atrophy of the technical arms of the department, it was difficult for requirement setters to get an unbiased view of the technological opportunities they needed. Moreover, new technologies, especially those related to networking, seemed to hold wonders for the battlefield of the future. Finding and fixing the enemy has long been the challenge of warfare. With advanced sensors, stealthy platforms, and global networks, warfare would be transformed. Senior policymakers' confidence in military officers' ability to make wise choices in this environment was not high. Contractors who had control of the technology seemed better placed for even this function. Handing the entire future requirements of a service or mission over to a contractor or a team of contractors, the essence of the Lead Systems Integrator model, was a comforting decision.

Since World War II, the trend in weapon system program design has moved toward greater and greater contractor responsibility. With the end of the Cold War, the government encouraged consolidation among firms in the defense industry. When combined with the responsibility shift that has been described, the government had unwittingly created a set of private arsenals, potentially as unresponsive as the public arsenals it replaced. The further evolution to the Lead Systems Integrator model, the result of the disintegration of the government's technological capabilities, may have sacrificed the political support that industry had offered weapon programs because of the opposition it generates in and out of government. Few want to trust the private sector with all public responsibilities, but the Lead Systems Integrator model seems to require it.

∞

The shift toward a dependency upon contractors that has occurred is understandable, given the difficulty in maintaining needed technical capacity within the civil service or the military. Initially the expected benefit seemed to be primarily technological. Later on, it seemed to be more promotional in the sense that contractors were more skilled at mobilizing political support for programs than were arsenals and shipyards that were subject to formal lobbying restrictions and had the comfort of assured budgets. The choices made by officials to seize the quick savings and increased political support that outsourcing provided are also understandable, given the existing pressures to focus on short-term results in government. But the total dependence upon contractors that has developed as a product of these decisions is unlikely to be either politically or technologically sustainable in the long run. Government officials cannot avoid being held responsible for the inevitable system failures—the huge cost overruns and missed time and performance targets—that such a dependency encourages. Some backtracking toward stronger public management of large-scale systems projects is highly probable.

COMPETITION AND INNOVATION UNDER COMPLEXITY

JEFFREY A. DREZNER

The products of the Department of Defense (DOD) acquisition process are perceived as becoming increasingly complex, emphasizing multifunction and multimission system configurations. Such weapon systems utilize network capabilities and systems-of-systems engineering and integration methodologies throughout their life cycles. The management and oversight of these complex programs have similarly become more complex. Changes may be needed in the organizations and procedures used to manage the development, production, and sustainment of these complex weapon systems.

This chapter discusses how "complexity" may affect the conditions under which competition and innovation yield the desired benefits.[1] Competition and innovation are not ends in themselves, but rather are a means to attain certain benefits in the context of weapon system design, development, production, and support. What are those benefits? What are the conditions under which competition and innovation yield the desired benefits? Have those conditions changed in ways that affect either the role of competition and innovation in defense programs or the benefits derived from that application?

The following discussion defines what is meant by "complexity" in the context of weapon system acquisition. It next describes the traditional view of competition and innovation in the acquisition environment prior to and through the 1990s. Given the changes commonly associated with complexity as defined here, the discussion then examines the implications for competition and innovation and ends by identifying implications for acquisition policy. This chapter draws substantially on past published work by RAND and others as well as unpublished work at RAND; useful references are listed at the chapter's end.

DEFINING COMPLEXITY IN DEFENSE ACQUISITION PROGRAMS

Before we can usefully discuss the implications of complexity for the use of competition and innovation in weapon system design, development, production, and support, we must first establish a working definition of complexity. In the context of DOD weapon systems, complexity can be thought of in three overarching dimensions—technical, organizational, and environmental. *Technical* complexity includes weapon system functionality and capability, including that related to the use of embedded information technology. *Organizational* complexity addresses the structures and interactions of the government and industry organizations responsible for system design, development, production, and support. *Environmental* complexity includes the political and economic context of the acquisition process, the threat environment, and the operational environment (how the systems are intended to be used). We expand on these three dimensions of complexity in the paragraphs that follow.

Weapon systems have become more complex over time. This is something of a truism and applies to the historical evolution of programs, not just to the more recent programs that have caught our attention. In general, new programs appear to be more complex than their immediate predecessors in terms of technology, functionality, and, perhaps to a lesser extent, their operational concept. Historically, this is the result of a natural evolution in which weapon designers and military users continually strive to improve and enhance warfighting capabilities. Under certain conditions, the use of competition stimulates innovation in weapon systems. Such an evolutionary pattern of improvement, whether derived from demand-pull or technology-push, applies equally to the commercial sector as well. It is the relative increase in complexity from one generation to the next that is of special interest. If each evolutionary step is relatively small, then management and oversight processes and practices will have time to adapt in parallel, and the required degree of adoption will be small. However, if the evolutionary step is large, there may be a significant mismatch between the complexity of the acquisition program and the institutional capacity to manage that program effectively.

Taken together, the three dimensions of complexity—technical, organizational, and environmental—suggest that we have entered an era in which the relative increase in complexity from the previous generation is fairly large.

The relative complexity of the weapon system itself is captured in technical complexity. Elements of technical complexity include the use of electronics, information technology, and software to provide critical functionality and capability beyond more traditional means. That these are increasing can be measured by the percent of acquisition program funds devoted to these technologies. These technologies reside in sensors, data processing, automation, communication, and data exchange. Many recent weapon systems are multifaceted, multifunction, multimission systems that include many more specific functions and performance capabilities than predecessor programs. Some programs, such as the first generation of semiautonomous unmanned air vehicles (UAVs) have no strong precedent and introduce entire new sets of capabilities.[2] Many recent programs also include the notion of "systems of systems" (SOS) in which many distinct systems are linked together through a common data network. In an SOS, each weapon system provides functionality by itself, but when linked together, the entire SOS provides capability that no single component system, nor all of those systems operating independently, could. The technical challenges in such complex systems emphasize systems engineering, software engineering, and system integration to a much higher degree than in the past. The Joint Strike Fighter (JSF, F-35), the Future Combat System (FCS), and DDG-1000 Zumwalt class destroyers are often cited as examples of complex systems. Such programs also tend to be fairly large (as measured by total program cost), which also makes them politically visible, adding an organizational dimension to complexity.

In a recent analysis, Robert A. Dietrick concluded that the complexity of weapon systems has been increasing over time.[3] He defines complexity in terms of the number of interactions among subsystems and the degree of integration of those subsystems, as well as the degree of integration at the component and part level—all aspects of technical complexity. Dietrick provides examples in aircraft avionics, airborne sensors, and computer processors; his definition of complexity is similar to what we mean by technical complexity. Further, Dietrick suggests that increased complexity—really increased functionality and capability—adversely affects program cost, schedule, and performance outcomes, though it is only one such factor.[4]

It is not just the weapon system itself that is complex, however. The second dimension of complexity concerns the organizations responsible for program management and program execution. Complex weapon system programs are managed by increasingly *complex organizations*. The relative increase in capabilities designed into modern systems requires increased breadth and depth of the government and industry workforce. The relatively large size (cost) of these programs adds an increased political dimension to program management. Large government program offices are staffed by a mix of military, civilian, and support contractors performing the full range of functions across a program's lifecycle. There are generally high levels of teaming among the industry components (at the prime contractor level and at lower tiers) because no single firm possesses the resources, capabilities, and political diversity required to fully execute the program itself. Government has increasingly relied on industry for both programmatic and technical capabilities, including program management, industrial base management, requirements formulation, systems engineering, and system integration. Officials of at least three programs—DD(X) (now DDG-1000), Deepwater, and FCS—have publicly stated that one reason they relied on industry for such important program management functions was due to a concern that the capabilities required to manage these complex systems did not exist in-house.

One consequence of complexity is the very large cost of complex systems. JSF, if it follows the current plan, will be the largest defense acquisition program ever executed, and FCS and the DDG-1000 Zumwalt class destroyer are in the same league. Expensive programs are politically visible and therefore vulnerable, which causes them to be managed with this in mind.

The lower industrial base tiers have become increasingly important as a source of innovation required to achieve program technical and system performance objectives. DOD policy, and economic policy more broadly, has often asserted that smaller firms are often more innovative. Mark Lorell has observed that it was often (though not exclusively) a smaller or second-tier firm that developed a key technological innovation leading to the next stage in the evolution of the U.S. combat aircraft industry.[5] Continued support of the Small Business Innovative

Research (SBIR) grant program also seems to support the notion that smaller firms located in the lower defense industry tiers are an important source of innovation. In most major defense acquisition programs, however, government-managed competition only occurs at the prime contractor level. Although the prime contractors might hold competitions among lower-tier firms for specific capability, the government may have little insight into these lower-level competitions, and little direct knowledge of the industrial base beyond the key second-tier firms involved in a program. Thus, the DOD has little information, and little ability to influence, competition in a portion of the market that may be an important source of innovation. As the top-tier firms focus more on system engineering and system integration functions, the lower tiers become an important source of technological innovation that is not being actively managed by DOD.

Finally, the complexity of the *acquisition environment* has increased. The threat environment is both broader and less predictable than in the past, resulting in increased complexity in terms of force and capability planning. The operational concepts of some complex systems are themselves complex in order to fully take advantage of new net-centric capabilities (e.g., FCS). Nontraditional or asymmetric warfare (e.g., counterinsurgency) introduces additional operational complexity. The complexity of the government and industry organizations and the rules governing them—statute, regulation, policy, processes—have also increased markedly.

These three dimensions of complexity—technical, organizational, and environmental—can be expected to affect the use of, and benefits from, competition in weapon system programs, including the resulting innovation attained though competition.

However, other factors affect competition and innovation that are not necessarily related to complexity, such as:

- Significant consolidation throughout the defense industry at all the tiers, but especially at the prime contractor level;
- Fewer and less frequent new program starts; and
- Large programs (e.g., JSF, FCS) that in the past would each have been multiple independent programs.

These trends and their implications need to be considered in any assessment of the effect of complexity on competition and innovation.

As we argue below, these noncomplexity trends may in fact dominate any effects on competition, while increased complexity has opened new areas to competition and innovation.

TRADITIONAL VIEWS OF COMPETITION AND INNOVATION

Competition and innovation are not ends in themselves, but rather are means to achieve certain goals.

Competition has long been a foundation of acquisition policy and contract awards for research and development (R&D), production, and services. In fact, there is a very strong bias in acquisition policy and federal regulation toward the use of competition, most recently illustrated by a policy directive from the under secretary of defense for acquisition, technology, and logistics.[6] In the defense acquisition context, we expect competition to provide lower prices, higher-quality products, cost control, improved efficiency, and innovation. In this sense, competition is sometimes thought of as a primary driver of innovation, though innovation may have other sources as well.

The conditions under which competition yields these benefits include the following:

- A large viable industry base, such that more than two firms or teams (with different firms) bid on a project. Viability includes both financial strength and a healthy and capable workforce.

- Some degree of industry or product sector maturity. If only the initial innovator plays, there is no competition.

- Product substitutability, which means that products are functionally similar across different firms.

- Many programs (i.e., frequent new starts) and a stable or growing budget. This condition is equivalent to a stable or growing demand function.

- Minimal barriers to entry. Such barriers might include capital equipment requirements or investment levels, workforce knowledge and skills, and even familiarity with government and DOD contracting and budgeting statutes and regulations, as these will affect program execution.

These conditions are part of the microeconomic model generally taught in undergraduate introductory economics classes. In particular, the plausibility of the "invisible hand" of a competitive market producing desirable outcomes depends on these and other conditions (e.g., free and full information). The lack of these conditions in particular defense sectors may prevent the expected benefits of competition from being realized.

It is important to note that an industry sector with only two firms and a government policy (implicit or explicit) to maintain the viability of both firms does not provide competition at the top tier (prime contractors). Although competitions can be held between teams led by these two different firms, each team knows at the outset that even if it loses, it will still receive a large enough portion of the program, or other programs, to remain viable. The industry base for large Navy surface combatants (Bath Iron Works and Northrop Grumman Ship Systems) and Navy submarines (Northrop Grumman Newport News and Electric Boat) are good examples of this challenge.[7] Both the DDG-1000 and Virginia class submarine programs have made preservation of the supporting industry base explicit goals of their acquisition strategies. As a result, the use of competition in these programs, and the benefits expected from competition, differ somewhat from the traditional.

Competition is thought of as a primary driver of innovation. However, competition is not sufficient in itself to generate innovation. Innovation depends on other factors as well, including funding levels; the existence of a core or "critical mass" of talent, capabilities, and resources in the same place at the same time (to include virtual colocation and other advanced collaborative tools in some cases); and a regulatory and institutional environment that encourages intelligent risk taking and out-of-the-box thinking.

Innovation is expected to result in new warfighting capabilities based on new concepts or technologies. Innovation is valued to the extent that it creates a warfighting competitive advantage between the United States and its adversaries. Innovation is also expected to be a primary source of a firm's competitiveness (thus coming full circle in this discussion). Beyond innovation of weapon systems or their use, innovation is also expected to result in improved business, design, development, production, and support processes (generally, increased efficiency).

Innovation arises from R&D investment, creativity, expertise, and sensing market trends. Technology-push and demand-pull both play roles in defense innovation. There are several frameworks that allow one to organize and think about the relationships between the factors affecting innovation. One such framework includes personnel capabilities and management, program management more generally (flexible vs. rules based), organization (institutional structure), technology, and workforce education and experience. These are not trivial factors: Defense Advanced Research Projects Agency (DARPA) was set up specifically to enhance innovation in defense-related technologies and concepts. With its highly educated workforce, flexible management, and relatively loose organizational structure, DARPA encourages out-of-the-box thinking. Its rules are set up to enable testing new concepts and technologies as quickly and inexpensively as possible. And DARPA has had many notable successes.[8]

Paul Bracken extends the work of two prior studies of innovation to develop a framework or model of innovation specific to the defense industry.[9] Six sets of factors are identified:

- National factors, which include education level, strength in science and technology, and supporting infrastructure (e.g., communication, transportation).
- R&D investment in a wide variety of projects, technologies, and sectors.
- Status and attractiveness of the sector (i.e., excitement and dynamism) as indicated by the degree to which industry in that sector is admired by consumers and students, the degree to which it is pushing the state of the art, and its ability to attract and retain top people.
- Competition in the sector, as determined by company strategies, industry structure, and rivalry.
- Demand conditions—in other words, the customer demanding capabilities requiring innovative new technologies.
- Related supporting industries including lower tiers and science and technology (S&T) base.

Note that competition is present in this model as a factor directly affecting innovation. The characteristics of the competition are important under this framework—that is, competition for ideas rather than cost or market share as the key driver of innovation in technology and product capability.

Additional factors affecting innovation or the conditions that facilitate innovation not explicitly identified in the models above include the following:

- An institutional and regulatory environment that encourages new concepts;
- Early adopters who are willing to buy and use initial versions of the innovation;
- A potential for significant demand for the product;
- High potential payoff; and
- Minimal barriers to entry.

A supportive institutional and regulatory environment is a critical foundation for innovation. An institutional structure that continually reinforces the status quo will hinder the ability of new concepts to be developed and tested. Feedback from early adopters is needed to help refine the product, demonstrate utility, and transition the innovation from the lab to a user community. In the past, the government has often been that earlier adopter. A large demand function establishes a potential market able to sustain enough sales to make the initial investment worthwhile. Since that investment entails risk, there must be a perception of a payoff commensurate with perceived risk, whether in terms of system performance, profit, or market share. Barriers to entry must be low enough to avoid seriously hindering the investment required for firms to establish a new market niche.

Industry sectors that are highly regulated tend to be relatively poor innovators. Increased formal rules and processes or large firms and bureaucracies may stifle innovation; there is less inherent flexibility, different expectations, and less openness to change. A tight regulatory structure and formal rules of behavior are thought to limit innovation (e.g., DARPA vs. DOD).

There is a set of assertions commonly made with respect to competition and innovation for which evidence is problematic. That does not mean that these assertions are incorrect, only that they are difficult to demonstrate with high confidence:

- Smaller, more flexible firms are more innovative. Some evidence supports this, though small firms often have difficulty finding the resources required to fully develop, test, market, and gain acceptance for a new concept or technology.

- Commercial firms are believed to be more innovative than the defense industry. This assertion underlies DOD policies concerning the use of commercial processes and products as well as alternative contracting strategies such as the "other transaction authority" (OTA) established to attract nontraditional firms to defense work.

- Innovation often comes from second tier or niche firms, not just the industry leaders. An interesting example of this phenomenon is in the military aircraft sector over the past 100 years: each new "technology era" in military aircraft (biplane, propeller monoplane, subsonic jet, supersonic jet, and stealth) was initiated by a second-tier aircraft firm or a niche firm (e.g., aircraft engines) that would then become a dominant player for that era.[10] To some degree, this assertion offers some support for the notion that smaller firms tend to be more innovative.[11]

HOW COMPLEXITY MIGHT AFFECT COMPETITION AND INNOVATION IN DEFENSE ACQUISITION PROGRAMS

Complexity itself has affected the nature of competition and innovation in the defense industry.

Many of the more recent programs are larger and more technically complex in terms of the use of information technology, system interdependence, and interoperability. Larger complex programs may require larger firms with substantial resources, breadth of capability, and the infrastructure to manage them effectively. Firms remaining in the defense market are relatively larger than they used to be and are themselves more complex (vertically and horizontally). The lead firm may focus more effort on system engineering/integration roles, including software development, rather than component and subsystem development and fabrication. In this sense, industry consolidation might be seen as an enabler for managing complexity.

The top-tier defense firms have restructured to better address technical, organizational, and environmental complexity. Technical complexity emphasizes systems and systems-of-systems engineering and integration, which in turn require an emphasis on this capability at the prime contractor level. Most of the top-tier defense firms have restructured in a way that reflects this focus, combining their military work under a new "integrated defense" business unit and hiring or training systems engineers. Boeing, Northrop, and Lockheed Martin all followed this pattern. These integrated defense business units also position the firms to better address interdependency and interoperability across system types, a challenge driven at least in part by technical, organizational, and environmental complexity. The emphasis on systems integration and system engineering capabilities offers a new niche for competition and innovation; the prime contractor competition in several recent programs—missile defense, FCS, DD(X)—emphasized systems engineering and integration explicitly.

This also elevates the role of the lower tiers; DOD-managed or influenced competition may now be more applicable and more important below the level of prime contractor. If DOD decides competitions at the prime contractor level because the government itself is unable to address the organizational complexity of a program, then competition at the lower tiers will be left to these large firms, who may decide such competitions based on different criteria than the government might prefer. At a minimum, increasing DOD awareness of the complete business base supporting a program may provide valuable information to policymakers on how competition can be applied in a particular case.

Complexity has also influenced the factors affecting innovation in many of the same ways. High barriers to entry remain, including capital investment and a workforce with the requisite characteristics. Complexity introduces yet another set of required workforce and organizational capabilities. There are many fewer firms at top industry tiers in mature industry sectors (e.g., fixed wing and rotary aircraft, large surface combatants, submarines, heavy armored vehicles). The government or defense-specific barriers to entry also remain, including knowledge and business processes that satisfy statutes and regulations as well as limited profit and limited growth in the defense sector.

An abundance of technical innovations (and associated concepts) has driven some of the complexity seen in today's acquisition programs. Complex systems have both advantages and disadvantages; they tend to be more costly, less reliable (more parts), harder to fix, and less predictable in behavior due to emergent properties. However, they also offer new capabilities useful to the warfighter.

But technical, organizational, and environmental complexity have also created new opportunities requiring substantial innovation in concepts and technology, leading to new capabilities and new niches within the defense industry. In more established sectors, innovation can be in the systems integration function, or in people, organizations, or management structures that bring diverse skill sets together. The potential of information technology to provide new capabilities or replace manned function with unmanned systems (e.g., automated fire control, shipboard firefighting, autonomous vehicles) has only just begun.

Although technical complexity dominates many discussions, it is not just technology that can be complex. Organizational and environmental complexities also offer opportunity for innovation. The changing nature of the threat has opened new sectors where less maturity gives innovation a relatively higher expected payoff. These capability areas include unmanned vehicles (air, ground, sea surface, and underwater), counterinsurgency (improvised explosive device, or IED, defeat, detection, communication/translation), space, and cyber warfare. Such new capabilities have implications for organizational structure of both the acquiring and user communities within DOD.

The technical, organizational, and environmental complexity discussed above may affect the conditions under which competition and innovation yield their expected benefits within the context of defense acquisition. However, there are other factors that also affect competition and innovation in defense programs, independent of complexity. There are relatively fewer new programs as compared to prior periods, at least in established defense sectors, reducing opportunities for competition (and innovation), but this was driven largely by budget pressure in the 1990s. There are fewer firms in the defense industry at all tiers, and, in some cases, very few firms are capable of designing, developing, and producing critical materials or components. Barriers to entry in the defense industry have always been high and are perhaps even higher now, at least at the top tiers. Workforce capability in the defense industry has also been identified as an issue; the older, experienced workforce is nearing retirement, and fewer younger workers are entering the defense industry. A scarcity of certain skills in the workforce can lessen a firm's ability to compete.

IMPLICATIONS FOR POLICYMAKERS

Complexity has contributed to changes in the nature of the weapon systems that DOD buys as well as changes in defense industry structure, how competition may be applied at the program level, the value of that competition, and the drivers of innovation. Policymakers should be aware of such changes when considering allocation of funds across possible weapon system investment portfolios, new program starts, acquisition strategies for programs, and management structure and processes.

Acquisition officials should consider the following observations when thinking about the application of competition to programs within an increasingly complex acquisition environment and implications for innovation:

- Little real competition currently exists in mature defense industry sectors. Complexity of programs or systems is only one cause. Other causes include fewer new programs providing opportunities for competition, an industry base that continues to consolidate in terms of the number of firms with specific capabilities, and increased teaming on large programs (i.e., spreading the business base).

- The globalization of the defense industry—an issue that has not been addressed in this chapter—offers some competitive opportunities by expanding both the number of programs and number of firms in the broader defense market. U.S. firms have competed in programs for other nations by offering versions of products sold to the U.S. military. Non-U.S. firms have competed in DOD programs either directly or by teaming with or acquiring U.S. firms (e.g., BAE and EADS). There are both near- and long-term impacts to globalization that warrant further study.

- Relatively new defense industry sectors such as unmanned vehicles offer opportunity for competition that can lead to innovation as well as provide other benefits expected from competition. These new sectors are expanding markets with lower barriers to entry and few truly dominant players.

- The organizations that manage complexity in weapon system programs are themselves complex. This applies to government and industry program offices as well as oversight organizations in the military services and the Office of the Secretary of Defense (OSD). In complex organizations, the interactions of many stakeholders can occasionally produce counterintuitive results.

- Government has traditionally focused competition at the prime contractor level. With competition among these large firms increasingly focused on system engineering and system integration functions, the competition that might produce technological innovations may more often happen at lower tiers. The government currently has few mechanisms to influence or manage competition among lower-tier firms.

- Bureaucracies tend not to innovate well, by their very nature. They are generally set up to ensure standardized processes rather than to develop new ideas. This characteristic applies to both government and the increasingly large defense industry firms in the top tier. In contrast, innovation seems to be facilitated by removing programs or projects from the mainstream. Examples include DARPA's accomplishments as well as the accomplishments of the several "rapid reaction" organizations set up to support warfighters in Iraq and Afghanistan. Historically, the relative success of classified (or "black") programs has been attributed in part to the nonstandard acquisition environment accorded them. Similarly, some large defense firms have set up advanced program operations to insulate them from the mainstream and foster innovation, such as Boeing's Phantom Works and Lockheed's Skunk Works.

One of the more important observations is that the factors affecting competition the most—fewer programs, budget pressure, industry consolidation—have little to do with complexity per se. Although complexity may change the nature of a competition by emphasizing large-scale systems engineering and integration rather than strict cost and performance variables, these other factors will still limit how competition can be applied in mature defense industry sectors. In contrast, complexity appears to have provided more opportunity for competition and innovation in relatively newer defense industry segments.

How can complexity in weapon system development be managed? There are two interrelated approaches; a mix of both is probably needed. One approach is to limit technical complexity in weapon system design by developing metrics for such complexity and using those metrics as part of the decision process when formulating a program's acquisition strategy. Such metrics might include the number of independent systems or large subsystems that need to be integrated, the number of interactions of systems within a weapon system, the number of external (or complementary) systems interactions required, and the number of organizations involved in design, development, and management.

A second approach is to adapt management techniques and institutional structures to better manage complexity. Hypotheses could be developed and tested at a smaller scale (e.g., program level) before applying them more widely. For instance, if technical complexity in a weapon system makes cost, schedule, and performance more difficult to predict, then an organization structured to respond to such uncertainties can be designed. Being responsive to uncertainty requires a good monitoring approach as well as considerable flexibility in making cost-performance trade-offs and allocating funds across a program. Pilot programs of the past have used this basic approach and have found some success—e.g., the initial JDAM (Joint Direct Attack Munition) pilot program or DARPA's Predator and HAE–UAV (high-altitude endurance unmanned aerial vehicle) programs. Simplifying decision processes may help minimize organizational complexity.

Policymakers should also acknowledge that the technical, organizational, and environmental complexity factors affecting acquisition suggest that we may not want to preserve the current government and industry structure; rather, we may want to consider how government can effect changes that respond to the evolving nature of acquisition and that create an environment that encourages innovation. Similarly, it is not clear that the current acquisition process needs to be maintained. Changes in the characteristics of what we buy and in the nature of the threat suggest a need for changes in the processes and institutional structures associated with acquisition. The policy levers that DOD has used in the past to shape industry, generate competition, and stimulate innovation are still relevant today and include the following:

- DOD is the only buyer for many new technologies. It can act as the early adopter for innovative concepts and technologies. DOD can use this status to shape R&D in the directions it wants to go.

- RDT&E (research, development, testing, and evaluation) funding—both the amount and distribution—is a major lever for DOD. DOD can diffuse private sector risk and ensure that a broad set of concepts and technologies are being pursued.

- The frequency and type of new programs, clearly related to funding amounts and distributions, are also critical. More programs provide more opportunity for competition and innovation. The increased use of smaller, focused concept and technology demonstration projects is an important policy lever. Advanced Technology Demonstrations (ATDs) and Advanced Concept and Technology Demonstrations (ACTDs) are examples of program structures whose use facilitates both competition and innovation. Careful attention must be paid to transitioning the results of such technology demonstration activities to major defense acquisition programs, particularly in terms of the doctrinal and sustainment issues often overlooked in technology demonstrations.

- Improved use of evolutionary acquisition strategies may also offer opportunities for competition and innovation. Such programs could be planned as a series of incrementally developed capabilities in which some portion of that incremental capability can be competed in an effort to encourage innovation.

- Use of less constrained contracting mechanisms, such as OTA, can attract nontraditional firms and allow the flexibility to both generate and pursue new ideas.

At the same time, however, it is important to recognize that current acquisition policy and practice, which have remain relatively unchanged for several decades, embody lessons in how to acquire complex systems and thus should not be discarded under the pretext of change without careful review.

NOTES

1. This work, which was sponsored by the Office of the Secretary of Defense under contract W74V8H-06-C-0002, reflects the views of the author. It does not reflect the views of RAND or any of its sponsors.

2. Robert S. Leonard and Jeffrey A. Drezner, *Innovative Development: Global Hawk and DarkStar—HAE UAV ACTD Program Description and Comparative Analysis*, MR-1474-AF (Santa Monica, Calif.: RAND Corporation, 2002).

3. Robert A. Dietrick, "Impact of Weapon System Complexity on Systems Acquisition," in James R. Rothenflue and Marsha J. Kwolek, *Streamlining DOD Acquisition: Balancing Schedule with Complexity* (Montgomery, Ala.: Center for Strategy and Technology, Air War College, Air University, Maxwell Air Force Base, September 2006).

4. Because technical complexity is very difficult to measure empirically, few analyses of program outcomes do more than simply raise the issue and assert a relationship.

5. Mark Lorell, *The U.S. Combat Aircraft Industry, 1909–2000: Structure, Competition, Innovation*, MR-1696-OSD (Santa Monica, Calif.: RAND Corporation, 2003).

6. John J. Young, Jr., under secretary of defense for acquisition, technology, and logistics, Memorandum, *Subject: Prototyping and Competition*, September 19, 2007. Competition has a long history in the U.S. defense industry. The very strong positive bias toward competition has its roots in the culture of capitalism and entrepreneurship that has driven much U.S. economic history.

7. Bath Iron Works and Electric Boat are both subsidiaries of General Dynamics. The various shipbuilding portions of Northrop Grumman have recently merged into a single entity. Thus, in some ways, there are really only two firms covering all large Navy shipbuilding programs—subs, surface combatants, carriers, and amphibious assault ships.

8. Richard H. Van Atta and Michael J. Lippitz, *Transformation and Transition: DARPA's Role in Fostering an Emerging Revolution in Military Affairs, Volume 1: Overall Assessment*, IDA Paper P-3698 (Alexandria, Va.: Institute for Defense Analyses, April 2003); Defense Advanced Projects Research Agency, *Technology Transition*, January 1997, http://www.darpa.mil/body/pdf/transition.pdf.

9. Paul Bracken, "Innovation and the U.S. Defense Industry," June 6, 2002 (unpublished input to RAND project). The two prior studies are Michael Porter, *The Competitive Advantage of Nations* (New York: Free Press, 1990), and Richard Nelson, ed., *National Innovation Systems* (New York: Oxford University Press, 1993). See also, John Birkler et al., *Competition and Innovation in the U.S. Fixed-Wing Military Aircraft Industry*, MR-1656-OSD (Santa Monica, Calif.: RAND Corporation, 2003).

10. Lorell, *The U.S. Combat Aircraft Industry*.

11. A similar analysis in other industry sectors has not been performed, so it is uncertain how widespread this pattern is.

REFERENCES

Adedeji, Adebayo, David Arthur, Eris Labs, Fran Lussier, and Robie Samanta-Roy. *NASA's Space Flight Operations Contract and Other Technologically Complex Government Activities Conducted by Contractors*, July 29, 2003, Congressional Budget Office.

Arena, Mark V., Robert S. Leonard, Sheila E. Murray, and Obaid Younossi. *Historical Cost Growth of Completed Weapon System Programs.* TR-343-AF. Santa Monica, Calif.: RAND Corporation, 2006.

Birkler, John, Giles Smith, Glenn A. Kent, and Robert V. Johnson. *An Acquisition Strategy, Process, and Organization for Innovative Systems.* MR-1098-OSD. Santa Monica, Calif.: RAND Corporation, 2000.

Birkler, John, Anthony G. Bower, Jeffrey A. Drezner, Gordon Lee, Mark Lorell, Giles Smith, Fred Timson, William P.G. Trimble, and Obaid Younossi. *Competition and Innovation in the U.S. Fixed-Wing Military Aircraft Industry.* MR-1656-OSD. Santa Monica, Calif.: RAND Corporation, 2003.

Bracken, Paul. "Innovation and the U.S. Defense Industry." June 6, 2002 (unpublished; input to RAND project).

———. "Innovation Systems in National Defense." April 25, 2002 (unpublished; input to RAND project).

Chao, Pierre, A., Guy Ben-Ari, Greg Sanders, David Scruggs, and Nicholas Wilson. *Structure and Dynamics of the U.S. Professional Services Industrial Base, 1991–2005.* Washington, D.C.: Center for Strategic and International Studies, May 2007.

Defense Advanced Projects Research Agency. *Technology Transition.* January 1997, http://www.darpa.mil/body/pdf/transition.pdf.

Dietrick, Robert A. "Impact of Weapon System Complexity on Systems Acquisition." Chapter 2 in James R. Rothenflue and Marsha J. Kwolek, *Streamlining DOD Acquisition: Balancing Schedule with Complexity.* Montgomery, Ala.: Center for Strategy and Technology, Air War College, Air University, Maxwell Air Force Base, September 2006.

Dombrowski, Peter J., Eugene Gholz, and Andrew L. Ross. *Military Transformation and the Defense Industry after Next: The Defense Industrial Implications of Network-Centric Warfare.* Newport, R.I.: Naval War College, September 2002.

Leonard, Robert S., and Jeffrey A. Drezner. *Innovative Development: Global Hawk and DarkStar—HAE UAV ACTD Program Description and Comparative Analysis.* MR-1474-AF. Santa Monica, Calif.: RAND Corporation, 2002.

Lorell, Mark. *The U.S. Combat Aircraft Industry, 1909–2000: Structure, Competition, Innovation.* MR-1696-OSD. Santa Monica, Calif.: RAND Corporation, 2003.

Schank, John F., Giles K. Smith, John Birkler, Brien Alkire, Michael Boito, Gordon Lee, Raj Raman, and John Ablard. *Acquisition and Competition Strategy Options for the DD(X): The U.S. Navy's 21st Century Destroyer.* MG-259/1-Navy. Santa Monica, Calif.: RAND Corporation, 2006.

Van Atta, Richard H., and Michael J. Lippitz, *Transformation and Transition: DARPA's Role in Fostering an Emerging Revolution in Military Affairs, Vol-*

ume 1: Overall Assessment. IDA Paper P-3698. Alexandria, Va.: Institute for Defense Analyses, April 2003.

Young, John J., Jr., under secretary of defense for acquisition, technology, and logistics. Memorandum. *Subject: Prototyping and Competition*, September 19, 2007.

SYSTEMS INTEGRATION FOR COMPLEX DEFENSE PROJECTS

EUGENE GHOLZ

Since World War II, the U.S. military has purchased defense *systems* rather than simple, one-off products. Systems are a combination of heterogeneous parts that work together to perform missions that could not be accomplished as well by simply operating the component parts independently. *Systems integration* is the process through which those systems are created. It includes the obvious concerns about connectivity among the components—managing the interfaces such that the various parts of the system can work together—but systems integration also goes beyond simple interoperability. It defines the characteristics of the system, starting with the analysis of alternatives to allocate tasks (which component will do what), specifies development objectives and technical requirements, and manages the various phases of the project (research, development, production, testing, and even disposal after the customer has finished using the system). Good systems integration allows the final product to serve the customer's true needs.

Although the U.S. military has long been very technology friendly, technology development is not one of its core skills. Especially as new defense systems rely more and more on electronics and software, individual platforms are becoming much more complex than in the past. The desire to increase network-centric capability, which makes previously separate platforms intimately depend on each other, reinforces this trend. Going forward, the military needs the means to make sure that the technological requirements derived from the wishes of its operational experts are attainable with a reasonable investment of time and resources.

The demand for systems integration, in quantity and quality, is at an all-time high. The military's buyers need to know what to buy, from whom to buy it, and what price to pay. To fill those needs, DOD and the services must have access to the core competency of organizations dedicated to specialized systems integration and technology management. Military-oriented systems-integration skill is based on advanced, interdisciplinary technical knowledge—sufficient understanding of all of the systems and subsystems to make optimizing trade-offs. It also requires a detailed grasp of military goals and operations as well as a reservoir of trust that bridges military, economic, and political interests. Some systems integration organizations also have some production capabilities (which may be either an advantage or a liability to the integration process), but systems integration is a separate task from platform building and from subsystem development and manufacturing.

This chapter explains what makes a systems integration organization good at its job and considers the strengths and weaknesses of various types of systems integration organizations (government, nonprofit, and private contractor).

WHAT IS IMPORTANT FOR GOOD SYSTEMS INTEGRATION?

An ideal systems integrator would combine a number of desirable characteristics, including technical awareness (with links to both academic research and the latest products in the marketplace), project management skill, customer understanding, organizational longevity (maintaining an institutional memory of past choices and investments and also promising reliable configuration control for the system's future trajectory), manufacturing expertise, and widely acknowledged organizational independence. This section explains each of these terms.

The bedrock of systems integration is familiarity with the technical state of the art in the wide range of disciplines that contribute to the components of the system. Systems integrators must be able to set reasonable, achievable goals for the developers and manufacturers of system components even as they "black box" the detailed design work for those components. To do that, they have to keep track of both academic research and the latest products to hit the marketplace, perhaps by scanning academic journals, new patents, and trade publications—and knowing how to evaluate what they see to distinguish marketing ploys and "vaporware" from real advances and useful products.

Many people who make good employees of systems integration organizations are the sort of people who appreciate and aspire to an academic affiliation—people with Ph.D.s who enjoyed their time doing research, playing with ideas, and solving problems. In fact, some experts have suggested that problem solving is the essential lesson for systems integration that Ph.D. scientists and engineers learn during graduate school: working in the lab to overcome all of the minor hiccups that would otherwise derail an experiment gives a certain mindset. But not every practical bench scientist makes a good systems integrator, either: many graduate research programs now emphasize very specialized knowledge of a particular experimental set-up, while the essence of systems integration is breadth of knowledge and curiosity to work with a range of technologies and disciplines. Universities are also a place to find curiosity and diversity of interests. So to get the best employees and to maintain technical awareness, systems integration organizations often benefit from formal (and informal) ties to nearby universities.

Technical awareness also helps systems integrators manage the design and production phases of a project. If one component maker has a problem that it can solve only at great expense but that could be solved much more easily by changing the requirements of a different component or by altering the interface standard in a way that would cost other component manufacturers less, the systems integrator is the one that must understand and implement the necessary trade-off among the various component specifications. The more the systems integrator knows about the subsystems, the better it will be able to perform.

Project management skill extends well beyond technical awareness, however—enough that it is really a separate qualification for systems integrators. Warfighters, technologists, and politicians all try to plan their expenditures, and as part of the budgeting process, they need project cost and schedule estimates that are as accurate as possible. For complex acquisitions with numerous, heterogeneous components, reliable estimates are difficult to come by, due to the vast amounts of information that must be managed to describe the current and projected state of progress. Systems integrators have to recognize both the state of the forest (the project as a whole) and of the trees (each component) so that they can redeploy resources to help keep a struggling part of a program on track or replace an underperforming subcontractor.

Overall, project management is much more complicated than simply following a program's technical progress and estimating its likely future trajectory. Project participants have incentives to hide certain information from oversight. Sometimes they believe setbacks to be temporary (that they will get back on schedule, the promised performance trajectory, or the estimated cost projection before they have to report problems); sometimes they fear that full disclosure will aid competitors or lead to pressure to renegotiate fees and expropriate profits. Managers also learn to phrase progress reports in favorable ways, or on very rare occasions, they submit false claims. Systems integrators have to see through all of these biases, understanding that often their decisions come down to judgment and organizational understanding rather than simply to technical analysis.

Good judgment depends on understanding the desired project outcomes—understanding what the customer really wants. That often depends on tacit knowledge and experience rather than formal, written statements. Defense acquisition projects have multiple goals, including maximizing warfighting performance, sustaining political support for the program and the national security strategy, and controlling spending on acquisition, maintenance, training, and operations. Major system acquisitions are complex even beyond their rising technical complexity.

Meanwhile, understanding what the customer wants contributes to projects in ways far beyond simple project management skill—to such an extent that it is really an independent desirable attribute of a good systems integration organization. Systems integrators have to make trade-offs and manage resources on projects in their customer's interests, even when the customer cannot clearly explain what he wants, certainly not in writing. At the same time, harkening back to the technical awareness ideal, these decisions also cannot ignore the interests and capabilities of the suppliers to such an extent that they either cannot produce the desired equipment or choose not to (which would often involve a complaint to Congress, creating a big headache for the military). Success depends on how deeply systems integrators understand their military customer's true needs and wishes—and how well the systems integrator can make the dreams mesh gently with reality.

Customer understanding is important for any organization, but it is especially difficult for defense systems integration organizations. The military is a complicated institution with a long history, unique traditions, and organizational biases developed over generations of operational experience. But customer understanding is also a moving target. A systems integrator must invest continuously in its military-operational knowledge base. It must monitor lessons learned from recent exercises and operational deployments, as well as changes in military doctrine and national grand strategy, to maintain the "right" kind of technical awareness. Ideally, systems integrator employees should participate in war games and exercises wherein the military tests new operational concepts and introduces virtual prototypes of future platforms and subsystems. Systems integration experts also can learn by deploying with warfighters to see their solutions in action. Working together in stressful environments helps personnel appreciate mutual idiosyncrasies.

But aggregating the experience from exercises and deployments requires another challenging step: real customer understanding requires interorganizational relationships that transcend individuals and projects. Military acquisition staffs are charged with thinking about the services' long-term needs rather than short-term operational pressures. A service's various acquisition projects interact with each other, and each project interfaces with many veto players within the service organization. A project can only stay healthy if its systems integrators understand the customer organization as a whole rather than just the preferences and style of even the most influential individual program officer.

Moreover, acquisition projects work on a very long time scale, reinforcing the need to understand the customer organization rather than the individuals working for the organization. The development phase alone for most major systems far outlasts a military officer's typical rotation in a program office. And after a system is deployed, it may stay in the inventory for decades, through which it needs maintenance and upgrades and, eventually, disposal. Through development, deployment, upgrades, and disposal, systems integrators are buffeted by their customer's demands—including not only the military customer's changing desires as its doctrine evolves and as it fights different enemies but also the rest of the customer environment including Congress's rising and falling interest in cost savings, environmental protection, and whatever else strikes the politicians' fancy. Systems integrators have to understand a lot and to continuously adapt.

A system's longevity also means that the program office needs an institutional memory, which is difficult to ensure due to the relatively rapid personnel turnover. And if the people in the program office have trouble providing that memory, so, too, do the defense industry primes and their subcontractors who build the equipment. Although most defense industry executives understand their firms' long-term interest in making a stable contribution to the national defense, their ability to work on a particular system only lasts until the end of their last contract—or until the congressional appropriation funding their work on the project runs out. The longevity of defense equipment is also a financial challenge, because firms cannot promise to stay out of bankruptcy for the life of the system. An ideal solution for systems integration has to find a way to remember what versions of a system exist and why, to maintain configuration control and documentation on how to use and improve each part of the system, to be able to explain a system's interface protocols so that it can work with yet-to-be-developed follow-on systems, and to keep track of what suppliers have been involved making components to ensure a reliable flow of spare parts.

Unfortunately, the best efforts to describe complex technologies—even those efforts aided by systematic processes like the Software Engineering Institute's Capability Maturity Models for systems integration—cannot capture all of the details that experienced engineers and technicians understand. People involved in actually making things have a certain feel for the limits of their capabilities, for what changes will be easy and what will cost a fortune, and for the effects of unpredicted or unexplained interactions within a system. The discipline of industrial engineering partly addresses these practical concerns, such as figuring out the most efficient assembly-line configuration to actually manufacture a product, but industrial engineers on an interdisciplinary systems integration team need experience, too. Furthermore, efficiency in the production process is not the only benefit that manufacturing experience brings to systems integration. Organizations that work entirely with abstract concepts or management expertise may lack a certain tacit "feel" for systems development and production that could lead them to propose impractical "solutions" that never quite work but nevertheless cost a lot of money to try.

Of course, a firm with production experience has an interest in continuing to get paid to produce what it is good at making—in fact, a very strong interest, since production is historically the part of the business in which defense companies make their profits. Asking a manufacturer for advice on trade-offs in designing a new system is likely to yield an answer that requires production techniques and equipment in which the firm has previously invested. There is nothing inherently pernicious about this tendency—customers also benefit when suppliers make what they are good at—but unconscious blinders or even an intentional effort to narrow the search for solutions in favor of the firm's experience will stifle creativity and may block the investment path that offers the most value for the consumer.

The high level of complexity involved in modern defense acquisition effectively guarantees that detailed decisions in systems integration will not be completely transparent to military customers, congressional appropriators, or the defense industry primes and subcontractors that supply components of the system. All of these groups must trust that the systems integrator has considered and protected their interests in making its decisions—that the systems integrator has enough organizational independence to decide impartially, weighing the concerns of customers and suppliers, including technical, political, and economic issues. Any organization that feels that its trust has been violated is in a position to create a scandal by complaining publicly. It is only constrained by the understanding that complaining too often or too loudly can subvert the entire process of providing for the national defense.

The best way to stave off scandal is to minimize systems integrators' interests in particular project outcomes, including both their pecuniary interests and their professional rewards. Integrators need to earn the same amount regardless of the choices they make about design trade-offs and project management, and they should receive accolades from their peers for making the right calls, whether tough or easy, rather than for making decisions that favor a particular technology, firm, or design philosophy. In the context of work for the government, the long-established standard in the United States is that organizations need to avoid not just actual conflicts of interest but the *appearance* of conflicts of interest. That standard especially makes sense in the context of systems integration for defense acquisition.

For practical purposes, we should not expect any single organization to be the best on all of these various criteria for systems integration performance: technical awareness, project management skill, customer understanding, organizational longevity, manufacturing expertise, and organizational independence. Perhaps it is even too much to ask that a single organization be the best at a majority of them—or that it even be *good* at all of them. Indeed, there are inherent trade-offs among some of these desiderata—for example, between manufacturing skill and perceived independence. So choosing the right systems integration organization for any given defense project—or even the right type of systems integration organization (in-house laboratory, specialized nonprofit organization, prime contractor, or another format)—is never easy.

TYPES OF ORGANIZATIONS FOR SYSTEMS INTEGRATION

Despite the increased attention brought by the prospect of an information technology-driven revolution in military affairs, systems integration is not a new concept for defense acquisition. Many organizations—including laboratories owned by the military services; private, for-profit contractors; and federally funded research and development centers (FFRDCs)—have long contributed to defense systems integration. Each type of organization has its strengths, and of course various organizations within each type are especially good in certain areas—for example among FFRDCs, MITRE tends to emphasize command and control work, while Aerospace Corporation tends to emphasize space systems. Thinking about the general performance of different classes of systems integration organization across the major categories of desired performance characteristics helps clarify the challenges of complex defense projects.

As the customer, the military should define projects' objectives, but the technical systems integration task is very difficult for the military itself to accomplish. The acquisition community focuses on government regulations and monitoring suppliers' compliance with cost, schedule, and other contractual terms. If project management were purely an administrative process, employees of the systems commands would have exactly the right skills, but project management also requires manipulation of vast quantities of technical data. Acquisition agents are usually not expert in state-of-the-art technologies and the innovative capabilities of various firms, so they work with other organizations that can supply systems integration capability and technical advice.

The military's systems commands still have their own laboratories designed to help with systems integration, and in principle the government could choose to expand its in-house systems integration skills. Indeed, some contemporary reform proposals promise to expand the government's contract management staff and expertise, and although many of the proposals speak generally of "oversight" capabilities, some clearly envision increasing the level of in-house technical systems integration knowledge, too.

Some of the advantages and disadvantages of in-house laboratories are quite straightforward. Because they are part of the military itself but are staffed by long-term civil servants, these labs have the potential for organizational longevity and customer understanding needed for good systems integration. On the other hand, the U.S. government has long been out of the business of making most equipment. Most arsenals have been phased out, and the Navy's shipyards emphasize repair and maintenance rather than design and construction. The general American distrust of government, especially government competition with private firms' activities, makes it unlikely that the government will regain manufacturing expertise in the future.

The difficult relationship between the systems commands' acquisition staff and the laboratories' technical personnel undermines the labs' effectiveness as systems integration providers. Scientists working in the labs often feel that the continuity of their research and their technical skills are undermined by frequent "cherry-picking" of researchers out of the laboratory and into the system command itself. For their part, systems command personnel tend to believe that scientists should support their immediate needs for advice on particular acquisition programs. Similarly, warfighters need quick fixes to get equipment working for an upcoming deployment, even if the solutions are temporary, nonsystematic, and nonrepeatable. Both of these pressures detract from in-house lab scientists' ability to pursue long-term research projects that contribute to their professional development. As a result, in-house laboratories cannot flourish on the criteria of technical awareness, even when they have unique equipment or capabilities that any systems development project in their area of expertise would have to use (e.g., a model basin).

The partnership between systems commands and in-house laboratories also has limited expertise in project management. Because laboratory scientists can resent the short-term demands of working with the systems commands, and the systems commands likewise worry about the quality of the technical inputs from their in-house laboratories, the two types of organization struggle to cooperate. Moreover, the political and organizational judgment required for good project management does not follow from a combination of acquisition policy expertise and technical awareness, no matter how good the partner organizations are at their core competencies.

But the laboratories' greatest weakness may be in the area of trust and organizational independence. Subordinate scientists (or even project managers in the systems commands) might not risk criticizing their bosses' preferred programs or the "pet rocks" of even higher-ranking officials—or at least outsiders might fear that this would be a problem. Insiders, on the other hand, see a different problem: civil servants working for military laboratories have a reputation for non-responsiveness within the military, and technically oriented military officers in particular face casual criticism that they are not "real" soldiers. In caricature, the military tends to fear that employees of the in-house laboratories are content to work slowly and steadily, going home to their families in the evening, while the warfighters face constant danger deployed far from home. The military lacks confidence in its own laboratories' responsiveness and customer understanding.

In sum, although in-house laboratories can offer some components of systems integration skills, they cannot currently offer the complete package. Expanding laboratories' capabilities would require major changes that are unlikely given the prevailing political and organizational cultures in American defense policy.

The second potential source of defense systems integration capability, for-profit prime contractors, in some ways simply reverses the advantages and disadvantages of the in-house military organizations. In recent years, for-profit contractors, whether individually or in partnerships, have led many of the cutting-edge military systems integration efforts. Some efforts have been more successful than others, but prime contractors clearly have a base of systems integration experience to build on. Their core business developing detailed designs and producing weapons platforms like tactical aircraft is related to the ever-more-complex task of integrating systems with multiple types of platform, but it is not the same thing.

Prime contractors combine a number of components of systems integration skill. Most directly, they are the only type of organization with substantial defense manufacturing experience. Although the firms often seem to fall victim to overoptimism about production cost and system performance when they are in the proposal phase for a major contract—and the defense contracting system encourages this overoptimism—the firms nevertheless often work through the difficult early phases of their contracts to deliver high-quality systems to the U.S. military. Moreover, the defense business is and will remain political. Defense contracts impose certain social goals on the defense industry labor force and supply chain—like cultivating small, minority-owned, or disadvantaged subcontractors—and the prime contractors' experience helps them to remain relatively efficient, given the constraints of their real-world environment. The ability to incorporate political constraints during design trade-offs is surely an important component of defense systems integration skill.

The experience working with diverse supplier networks while planning and executing contracts to deliver weapons platforms also gives for-profit prime contractors a relatively strong position in project management skill. True, the traditional prime contractors sometimes struggle to keep projects on schedule and within budget limits—even as complex commercial ventures like Boeing's 787 jet also sometimes face project management snafus. As projects grow in complexity and scope, no amount of project management experience and adept organizational routines is likely to meet all of the projects' goals. But considering the difficulty of the project management task involved in most major defense acquisitions, the for-profit prime contractors are well positioned to be effective project managers.

Third, prime contractors' general level of customer understanding tends to be very high. It is precisely their ability to understand military jargon and to track various military ideas and doctrinal initiatives that distinguishes defense-oriented companies from other high-tech firms that focus on large commercial customers. Prime contractors' strategic planning departments hire retired military officers. Although the value of any insider knowledge that might arrive through the revolving door fades relatively quickly as programs evolve, the general "feel" for military organizations and the ability to navigate the complex customer organizations persist.

Finally, private firms are largely exempt from civil service rules, allowing them the flexibility to hire top technical talent when necessary. Private firms can also offer stock options to scientists who crave equity compensation, and they have incentives to support technical teams' internal rapport and to invest in human capital improvement. Managing technical personnel is a core competency of technology-dependent private firms, including defense industry prime contractors.

On the other hand, prime contractors' opportunities for high-level systems integration are hampered by weaknesses in a couple of other areas. As discussed previously, for-profit private firms struggle to promise the kind of organizational longevity and focus that some systems integration tasks require. Even if the government were to offer a long-term contract to one of the for-profit primes, promising to keep them in the systems integration business for the duration, shareholders' pressure for constant corporate growth and top design teams' interest in pursuing new ideas would drive firms to transition engineers to new development projects.

But the major prime contractors really fall down on the potential for conflicts of interest or at least for the *appearance* of conflicts of interest. Military and congressional customers might reasonably fear that a manufacturer's trade-off analysis might be biased in favor of the sort of alternatives that the manufacturer is good at making. More subtly, the production contractors' technical understanding of particular systems and solutions might unintentionally skew the prime's analysis of alternatives. Historically, competitors of the prime contractor that wins a contract fear that the winner will gain insider access that will help it in future competitions, and they complain about the potential unfairness. Other firms in the defense industry also fear that they will have to explain their best intellectual property to the prime if they want to compete for subcontracts. This fear again leads them to complain about unfairness or, perhaps even worse, leads them not to offer their best ideas for components, undermining the quality of the final product.

Past efforts to design contracts to avoid these pitfalls, such as building firewalls within firms or requiring the primes to create independent subsidiaries to manage each project, have generally failed. Confidential documents seem to have leaked across internal firewalls with some regularity, so the contractual terms have done little to dampen complaints about conflicts of interest. But the legal barriers do inhibit the prime systems integrator's access to the firm's high level of technical awareness, customer understanding, and manufacturing experience, eliminating the very advantages that attract the military to giving systems integration contracts to primes.

During the early days of the Cold War, the difficulties with using prime contractors for systems integration led to the creation of another type of institution—the FFRDCs—to work on major defense contracts. FFRDCs are specially chartered nonprofit corporations that receive long-term government contracts (typically five years at a time) that are not tied to work on particular projects. Instead, the FFRDCs are paid to do a set amount of work each year, and various project offices (and other government organizations) ask the sponsoring agency to allocate FFRDC work effort to their needs. FFRDCs cannot compete for government production contracts, although it is sometimes difficult to draw the line between prohibited production and some kinds of allowable research, development, and prototype development.

Like in-house laboratories and prime contractors, FFRDCs have advantages and disadvantages as systems integration service providers. The historical strength of FFRDCs has been their reputation for high-quality, objective advice. Through flexibility in salary negotiations and their quasi-academic status, FFRDCs have been able to attract high-quality personnel. Many also have formal ties to leading universities, and the rest have dense informal ties. Some research-oriented FFRDCs are in many ways like universities without students, allowing employees to focus on research without as many administrative or teaching-related demands. And the FFRDCs' long-term contracts reduce the speed of employee turnover, contributing to the institutions' historical memory and ability to promise steady configuration-control procedures.

At the same time, the FFRDCs' promise not to compete for production contracts and to provide equal access to all contractors while safeguarding proprietary information has with a few notable exceptions allowed them to avoid allegations of conflicts of interest. Meanwhile, their close relationship with their sponsoring agencies, whether the Office of the Secretary of Defense or one of the military services, gives FFRDCs a very strong customer understanding. In many cases, they were founded as sounding boards and advisory organizations for the leadership of the defense establishment, and their funders generally trust that the FFRDCs have the customers' interest in mind essentially all the time. On the other hand, individual FFRDC employees do not draw their salary from particular project offices, and their promotion prospects do not depend on satisfying the military chain of command.

Thus, the FFRDCs, though unable to criticize their customers too much or too publicly, are also paid for their independence of thought, which protects them as long as they stay within certain informal but well-known boundaries.

However, despite their dramatic successes early in the Cold War, FFRDCs are widely accused of having slipped on project management skill, and perhaps their lack of manufacturing experience has taken a toll on the practicality and affordability of some of their recommendations. Although leaders of FFRDCs claim that their nonprofit status allows them to charge less than a hypothetical for-profit adviser with equivalent technical skills, many critics allege that the lack of a profit motive in FFRDC work leads to inefficiency and the potential for featherbedding. Each of the current line-up of 10 Department of Defense FFRDCs has been around for decades, and their long-term contracts and the low risk of replacement by a competing new FFRDC may have rendered them less technically dynamic organizations than they were in the past. Each has a set of "hero" programs on which it cut its teeth years ago—a legacy that it now has an interest in preserving and defending.

For-profit firms that agreed not to engage in any production might be able to offer the benefits of FFRDCs while avoiding the controversies linked to nonprofit status. However, it is difficult to imagine such a firm nurturing a major laboratory with an independent research capability and agenda—necessary to maintain top-level systems integration skills—without competing for production contracts that would cover the overhead cost and appeal to shareholders' desire for growth. Furthermore, long-term contracts might mute the supposed competitive benefit of for-profit competition. Because the long-term contracts are essential for avoiding pressures for bias and for capturing the advantages of organizational longevity, perhaps the best solution to reduce lax performance is oversight rather than competition in the systems integration niche.

Legislation currently limits the budgetary resources available to FFRDCs, and for a number of years, a law prevented the establishment of any new FFRDCs. In 2004, the high level of perceived threat of terrorist attacks against the United States overcame congressional reluctance, and Congress created the Homeland Security Institute to support the Department of Homeland Security. But aggressive competition from

Table 5.1. Summary of Systems Integration Capabilities by Organizational Type

Key Attributes for Systems Integrator	Systems Integration Model		
	Government Laboratory	Industry	FFRDC
Technical awareness	–	+	+
Project management skill	–	+	+/–
Customer understanding	+/–	+	+
Organizational longevity	+	–	+
Manufacturing expertise	–	+	–
Organizational independence	–	–	+

for-profit advisers and general disquiet with set-asides, especially for "elite" institutions, constrain the potential expansion of FFRDCs to provide additional defense systems integration.

Table 5.1 summarizes the systems integration performance attributes of the three major types of systems integration organizations. The table indicates the relative performance of the organizations with a very blunt measure: + for strong baseline performance, – for weakness on a particular characteristic, and +/– for an organizational type with mixed capabilities in a certain area of systems integration. The table also only reflects the capabilities of the types of organizations rather than of individual laboratories, prime contractors, or FFRDCs. Nevertheless, the summary may prove useful as a quick view of the landscape of systems integration for complex defense acquisition.

CONCLUSION

As systems integration becomes ever more important in the face of increasing complexity of military systems, the search for better institutional arrangements will only intensify. Unfortunately, systems integration itself is a complex task best performed by an organization with a host of qualities, and as in many parts of life, all good things for systems integration do not go together. Different projects will emphasize different qualities: technical awareness, project management skill, customer understanding, organizational longevity, manufacturing expertise, and organizational independence. But it will rarely be obvious, especially in advance, which of those qualities is most important at a particular time and place.

The process of systems integration is about making complex tradeoffs within a project, and we should not expect complex projects to always work out as planned, regardless of the systems integration organization employed. Similarly, we should not expect to be able to assign the systems integration task to the "right" organization, whether an in-house laboratory, a prime contractor in the defense industry, or a specialized nonprofit FFRDC. But we should strive to think carefully about the advantages and disadvantages of each type of organization and to seek a balance in the overall systems integration capabilities available to American defense policymakers.

A NEW WAY OF THINKING ABOUT ENTERPRISE CAPABILITY DEVELOPMENT
NETWORK-CENTRIC, ENTERPRISE-WIDE
SYSTEM-OF-SYSTEMS ENGINEERING

JEREMY M. KAPLAN

This chapter describes the fundamental challenges that large enterprises face in developing large numbers of interacting systems in a complex world and addresses these challenges from integrated social, organizational, and technical perspectives.[1] It proposes and develops the concepts of network-centric, enterprise-wide system-of-systems engineering as a better way of addressing those challenges. It proposes guiding principles, enabling concepts, and specific recommendations for DOD.

FUNDAMENTAL CHALLENGES
In a complex and rapidly evolving world, the missions and challenges of a large enterprise may change rapidly as its competitive environment evolves. There may be significant uncertainty and debate within the enterprise over the relative importance of different missions, the capabilities required, and the number, kind, and features of the systems needed to create these capabilities. Developmental speed and efficiency require the enterprise to organize so that large numbers of systems can be developed simultaneously and continuously—but how can these multiple efforts be coordinated? In addition, new technologies open new possibilities, but exacerbate integration problems among legacy systems and those under development.

The fundamental challenge facing large enterprises in developing capabilities under these circumstances is how best to continuously develop large numbers (perhaps hundreds or thousands) of different systems to create capabilities to meet multiple, uncertain, and evolving needs. Individual systems are built and "optimized" by quasi-independent developers under different governance structures, to different and evolving sets of requirements, to provide different but interacting capabilities to support multiple but evolving missions. Nevertheless:

- The systems must become and remain suitably interoperable with each other and with existing systems.

- The overall system-of-systems must be "best" in some sense that considers overall multi-mission capabilities, agility, performance, cost, and risk.

The enterprise and its elements must address the specific challenges of how to

- best allocate resources,

- develop and coordinate required capabilities,

- coordinate and manage developmental efforts, and

- achieve interoperability while encouraging experimentation and initiative.

These challenges exist at the enterprise level and at many intermediate (and not necessarily hierarchically structured) levels—e.g., military service, joint, and functional—across the enterprise. And "best" may be different for different systems under different circumstances—perhaps emphasizing high performance for certain functions and low cost for others.

How can this fundamental problem be addressed? Complexity and change preclude overall optimization. Loose central control causes significant interoperability problems and poor mission performance as each system goes its own way. Top-down direction does not work. Specification of the features (or even the interoperability standards) to be developed in hundreds or even thousands of appropriately interoperable systems (or net-centric services) freezes innovation. Tight central control slows issue resolution, inhibits inventiveness, and guarantees technologically obsolete systems.

How can we simultaneously define needed capabilities, allocate resources, and develop systems when decisionmaking in each of these processes cascades down through multiple organizational layers? How can we encourage experimentation and initiative while still achieving interoperability, and how can we continue to innovate and still maintain interoperability with past systems still in use?

CHARACTERISTICS AND UNDERLYING PROBLEMS OF A LARGE ENTERPRISE

To make progress, we must examine how the underlying problems of large, complex enterprises develop. Large enterprises with single missions can usually solve their problems through the development of systems, or systems-of-systems (SOSs), by using classical systems engineering—requirements analysis, functional decomposition, and synthesis. The space program of the 1960s provides an outstanding example of this. However, as an enterprise grows and develops large numbers of missions that require large numbers of systems and as the environment becomes complex, efficiency of governance requires the organization to develop suborganizations (e.g., Army, Navy, Air Force). Complex missions then require these to develop systems-of-systems, and crosscutting missions (e.g., communications, intelligence, etc.) require the development of crosscutting de facto or de jure systems-of-systems. The organizations that develop these separate systems-of-systems develop their own cultures (e.g., the separate military service cultures), so that apportioning resources becomes a challenge, and to resolve conflicts it often becomes essential to separate requirements development from resource allocation from systems development/acquisition. Problems then develop within and across these multiple systems-of-systems and governance processes.

The complexity of the governance situation in DOD is astonishing. Each military service and several defense agencies and commands develop systems. Where appropriate to improve capabilities, save resources, or develop more efficiently, these are organized into de facto and de jure systems-of-systems (via program executive officers, or PEOs, portfolios, or functional oversight). Other organizations (e.g., Joint Staff JCIDS [Joint Capabilities Integration and Development System] Functional Capabilities Boards and Office of the Secretary of Defense principal staff assistants, or OSD PSAs) attempt to define needed capabilities and coordinate capabilities and system development across services and sometimes across functional areas. Resource allocation may cut across these processes to varying degrees and at various levels, but usually remains within military service boundaries. This leads to the system-of systems environmental characteristics of the left column of figure 6.1.

First and perhaps most significant, systems-of-systems, or SOSs, are subject to complex and overlapping governance. They may have independent requirements development, resource allocation, and acquisition processes, even within a military service. The same system may be subject to multiple governance authorities (e.g., an Army communications system may be subject to proposed trade-offs by both the Army and the assistant secretary of defense for networks and information integration, or ASD/NII).

Each oversight organization and each system developer is trying to do the best it can under the guidance it has, but guidance, which is developed by different groups at different times, takes time to cascade down multiple levels in each organization and is constantly changing as military needs change, problem understanding improves, and new technologies are developed.

The size of the enterprise may present problems for the movement of developmental information across the enterprise. Because information-sharing requirements (both for systems development and for operations) may be uncertain and changing, system developers may be unaware of all the systems that may depend upon them, or that they might be able to draw on, or that their users may need to collaborate with. The enterprise may grow so large (and DOD has) that multiple systems with similar or overlapping capabilities may be developed by organizations that are unaware of, or have no time to deal with, each other.

Many SOSs (e.g., land combat systems and communications systems) have indefinite or infinite lifetimes (that is, there will always be a need for the capability), so that the problem of interacting with legacy systems will never go away and the systems currently under development will someday become some newer system's legacy.

Finally, SOSs are fielded to perform complex missions in coordination with other systems and systems-of-systems against opponents who are evolving their tactics and systems. Thus the analytic problems associated with optimization are complex and are rarely amenable to exact solution in time to influence a decision.

These characteristics lead to a common set of underlying problems for individual SOSs and the enterprise. Complex, independent governance and the size of the enterprise lead to developmental friction—

Figure 6.1. Conceptual Framework: Characteristics, Problems, Net-Centric Guiding Principles, and Enabling Concepts

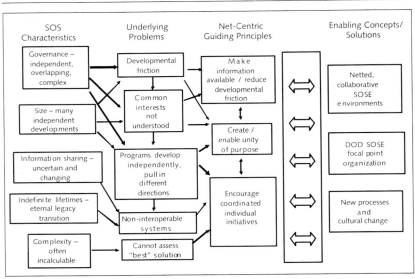

the wasting of energy in trying to coordinate system development. So much energy is expended in information discovery, against bureaucratic inertia, and in attempts to coordinate that individuals at the working level eventually stop trying, and common mission and developmental interests are not uncovered or understood. Programs thus develop independently, and their systems become non-interoperable.

These problems are so universal that DOD has evolved a name—"stovepipes"—for independent systems-of-systems, and the hardworking and well-meaning people who have developed them recognize the profound cultural causes that prevent the creation of the tools and processes needed for the technical integration of these systems.

Systems-of-systems have been created in DOD at many levels. The military services have created independent functional systems-of-systems for command, control, and communications (e.g., the Army's LandWarNet, the Navy's ForceNet, and the Air Force's Constellation-Net). At the next level up, and cutting across the military services, DOD has created joint systems of systems (e.g., TRANSCOM's integrated logistics system), and defense agencies (such as the Defense Information Systems Agency, or DISA) have created joint command, control, and communications (e.g., Global Command and Control

System, or GCCS; Global Combat Support System, or GCSS; and the Defense Information Systems Network, or DISN). At a higher level, DOD is still struggling to define the boundaries and governance of the entire Global Information Grid (GIG) system-of-systems and the nested systems-of-systems that form it.

These systems-of-systems are created whenever a controlling organization believes that better performance will be achieved through more centralized management. The systems-of-systems may be nested and may interact in complex ways. However they are defined and created, they have some features in common: they must always interact with multiple systems and systems-of-systems that they do not control; and as they grow larger, better interoperability is achieved at the expense of slower individual system development and lower individual system performance. This latter point is extremely significant. Too much control at too high a level can slow system development and diminish overall performance, as systems developers are required to spend too much time and energy applying for exemptions to rules they find overly constraining.

NETWORK-CENTRIC, ENTERPRISE-WIDE SYSTEM-OF-SYSTEMS ENGINEERING

Systems engineering is the discipline of developing systems optimized to solve specific problems. Its top-down, structured approach is of limited utility in organizing large numbers of people to tackle rapidly evolving problems in the complex context previously described. It is more fruitful to understand the SOS problem and to develop principles and concepts for tackling it.

This section explores some guiding principles and key enabling concepts for system-of-systems engineering. Of course, systems-of-systems may exist at multiple levels in large enterprises, and the specific challenges and specific solutions applied by the system-of-systems engineer will vary with the level at which the SOS is developed. The SOSs we are concerned with are generally large, enduring collections of interdependent systems under development over time by multiple independent (or semi-independent) authorities to provide multiple, interdependent capabilities to support multiple, complex missions.

System-of-systems engineering, as defined in this chapter, is the cross-system and cross-community discipline that enables the development and evolution of mission-oriented capabilities to meet multiple stakeholders' evolving needs across periods of time that exceed the lifetimes of individual systems. It does not replace systems engineering, but adds to it.

It is important to note that the system-of-systems engineering does not constitute "a solution" to the enterprise's problems, because these problems stem from uncertain boundedness in multiple dimensions (mission, physical, and logical boundaries, lifetime, governance, and information flows) and do not have "a solution." But these principles and enabling concepts greatly improve the ability of the enterprise to develop capabilities and systems in a complex, multi-mission environment.

Because all enterprise developmental processes (e.g., requirements development, resource allocation, and systems development) have important common characteristics and because we want an approach that works with current and potential future governance processes, we begin by generalizing developmental processes and putting them on a common footing with a common vocabulary.

We define a system-of-systems authority (SOSA) to be any organization that has authority (e.g., oversight, resource allocation, requirements definition, certification) derived from any legal source. So, PEOs, PMs (program managers), and OSD PSAs are all SOSAs, as are Joint Staff JCIDS Functional Capabilities Boards. Each SOSA is responsible for creating a "best" system-of-systems in some sense—perhaps the greatest performance for some mix of missions, good performance for the greatest mix of missions (agility), or the lowest-cost solution for some well-understood function. Each SOSA thus has systems issues to address and needs analytical support by someone—its system-of-systems engineer (SOSE)—who can address and coordinate the addressing of these issues.

A system-of-systems engineer, or SOSE, requires and reports to a system-of-systems authority (SOSA) and provides technical analysis, coordination, and support. The SOSEs must work through their SOSAs and not create a competing governance structure. Each SOSA retains its original authority (i.e., oversight, resource control, etc.) so that the existing governance structure of the enterprise is not affected. The SOSE's goals are the development and evolution of the best (in a sense decided by the SOSA) overall mission-oriented capabilities.

This SOSA/SOSE relationship is essential. SOSEs who have been set up independently or without adequate backing have been resisted or sabotaged and their processes and recommendations ignored.

The SOSE should perform three major roles:

- The SOSE should provide and coordinate overall analytical support for its SOSA. This classical systems analysis and systems engineering role enables the SOSA to both improve its system-of-systems and obtain support for it at higher levels. Of course, this SOSE role includes providing technical vision, technical assessments, and technical/cost trade-offs for the SOSA.

- The SOSE should create the environment in which individual systems developers can innovate and perform trade-offs within their own systems in the context of the full system-of-systems, so that they can investigate and implement changes to their systems that improve the overall system-of-systems in the sense the SOSA desires. This role may include the development and promulgation of technical architecture, a performance modeling framework, and even systems engineering guidelines. It is characterized by the goal of unleashing the fullest potential of each individual systems engineer toward improving the SOSE's entire system-of-systems and not just the system engineer's individual system.

- The SOSE should work with and coordinate with the external systems and systems-of-systems that the SOSA's system-of-systems must function with in its broader operational contexts. Thus a communications SOSE might work with a weapons system SOSE and a logistics SOSE to enable the development of better capabilities for some broader set of missions and functions.

These three roles are essential. Some current SOSEs (usually Lead Systems Integrators hired from industry) focus primarily on the first. They are more inclined, after the initial exploration of solutions, to perform overall analyses and flow requirements, top-down, to the individual systems. The lack of attention to the third role (coordinating with other systems-of-systems) results in poorer overall mission performance for the department as a whole. The inadequate focus on the second (creating an innovative environment across the individual systems) results in poorer evolving performance of the SOS under their purview.

Figure 6.2 illustrates the roles and relationships of the SOSEs and SOSAs in an enterprise context. The SOSAs in the figure may oversee SOSs on any scale from major complex systems to SOSs that include

Figure 6.2. Relationships of the SOSA and SOSE

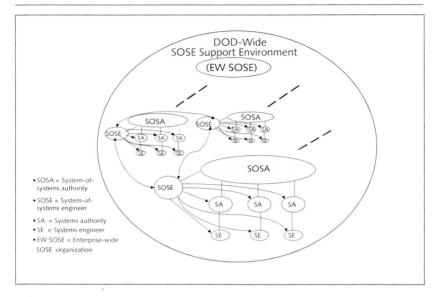

other SOSs. Intermediate-level SOSs (such as a communications system-of-systems that encompasses military service and DISA-owned SOSs) may be included. The systems authorities (SAs) in the figure may represent any organizational authority subordinate to the SOSA and responsible for a single system—e.g., program managers if the SOSA is a program executive office involved in systems acquisition.

These concepts lead to flexible and scalable process involving self-synchronization. SOSAs and SOSEs can self-organize to solve mission-oriented problems within their SOS (among their systems) and among themselves (across their SOSs).

The relative emphasis that an SOSE should place on its three roles changes as the underlying SOS becomes larger and contains more nested SOSs. An SOSE for a small SOS should concentrate on its classical systems analysis and systems engineering role, with some emphasis on internal and external coordination. For a larger SOS, the coordination role among systems should receive far more emphasis, because increased complexity requires greater coordinated analytical efforts on the part of the individual systems engineers. At the enterprise-wide level, the SOSE's coordination role should receive almost the entire emphasis for a very fundamental reason: the technical and other issues across the entire enterprise are so large and complex, and the governance processes of large enterprises so independent, that any attempt by an enterprise-wide SOSE to dictate solutions is doomed to resistance and failure. Government by the consent of the governed is essential if any progress is to be made. Thus the role of the enterprise-wide SOSE organization is to foster the environment, processes, and products to facilitate the work of the SOSEs and create excellence in SOS engineering.

To address the underlying problems previously described (see figure 6.1), SOSAs and SOSEs must be guided by the principles of enhancing unity of purpose, improving information availability, and encouraging coordinated individual initiatives.

Unity of purpose is W. Edward Deming's concept that when workers understand the underlying mission, they are motivated to improve it. Information availability attacks the problem of developmental friction brought about by independent governance and large size. Coordinated individual initiative is the active principle of encouraging (rather than controlling) initiative at all levels as long as it is consistent with the overall mission.

The first guiding principle is to create and enable unity of purpose. In general, SOSAs create unity of purpose, but SOSEs enable its implementation with architectures, tools, and developmental environments. The better SOSAs, SOSEs, and especially individual systems engineers understand their overall purpose and possess the tools that enable them to see how changes in their systems can further that overall purpose, the better they are able to propose and implement changes that improve overall enterprise capabilities.

The second guiding principle is to reduce developmental friction by making information freely available across the enterprise. The more effortlessly each coordinating and developmental activity can see all the factors that may affect it (or that it may affect) and its components, the better it is able to influence and adjust to produce a more effective enterprise. This must be done ruthlessly, because supplying information normally requires effort and entails risk.

The third principle is to encourage individual initiatives to the maximum extent possible, consistent with coordination toward achieving overall enterprise capabilities. Many of the enabling concepts that follow could increase overall central control, a temptation that must be resisted. The enabling concepts and tools must be used to create a context in which initiative can be taken at all levels. This is true both across systems-of-systems and within them. Senior management must surrender tight control in favor of loose coordination, especially at the initial conceptual and developmental stages.

This approach to SOS engineering can be thought of as an extension of net-centricity to the broader developmental community. Net-centricity posits that organizations are more effective when they bring "power to the edge"—that is, when they make information freely available to those who need it and permit free collaboration among those who are affected by or can contribute to a mission. Benefits of net-centricity in an operational setting include better situational awareness and problem understanding; better development of goals and objectives; better communication and understanding of commander's intent; and better planning, collaboration, and self-synchronization in pursuit of solutions. Net-centric SOS engineering creates processes and tools that enable net-centric culture and bring the benefits of net-centricity to the broad community that sets requirements, allocates resources, and develops systems. It enables the multiple activities of systems engineers and SOSEs to go on simultaneously and cooperatively.

ENABLING CONCEPTS AND SPECIFIC RECOMMENDATIONS FOR THE DOD GLOBAL INFORMATION GRID

The three major enabling concepts of enterprise-wide systems-of-systems engineering are networked collaborative environments, a network-centric culture, and the enterprise-wide SOSE organization. These solution concepts can connect systems developers involved in all three developmental processes, thus enabling collaboration among the full collection of people who define requirements, allocate resources, and develop systems. The concepts must be implemented to accord with and to further the network-centric guiding principles previously described or they are unlikely to produce much benefit. In addition, the networked collaborative environments, cultural changes, and enterprise-wide SOSE organization are mutually dependent solutions in the sense that none will be very effective without the others.

COLLABORATIVE ENVIRONMENT

Figure 6.3 provides specific examples of potential components of a networked collaborative systems engineering environment that can

Figure 6.3. Netted Collaborative Engineering Environment Recommendations

- Visibility across DOD
  - Minimum posting requirements
  - Joint systems/services architecture
  - Joint operational architecture
  - COI data repository
  - Future interoperability technologies
  - Productivity/posting software
  - Dependency tracking software
- Modeling and simulation
  - User's forum
  - Modeling framework
  - Model development tools
- Contextual design and development
  - Joint distributed development & test environments to enable the resolution of interoperability and performance issues
- Guidance for DOD
  - Open interoperability standards
  - Commercial participation
  - Reenergize activities
  - Enterprise services
  - Mandated use
  - Integrated operational management (NETOPS)
  - Implementation guidance for systems engineers

greatly improve systems development across DOD. This chapter will only briefly describe some of these components. A fuller description of them and of the concepts in this chapter can be found elsewhere.[2]

The vision of collaborative systems-of-systems engineering environments is one where the entire enterprise, including individual communities of interest, are fully networked together, and each is enabled with needed collaborative tools, analytical capabilities, and effortless access to information.

The utility of networked collaborative environments comes from their ability to empower developmental activities at all levels from individual system developments to major SOS development. Features of this collaborative environment would include, but not be limited to, data repositories and Web sites for posting and sharing information, integrated and analytical tools and modeling and simulation environments, integrated development and test environments, and an approach to guidance that recognizes the implicit tax that guidance imposes.

The purpose of improving visibility across DOD is to eliminate developmental friction. Individual system developments must know about related systems without engaging in time-consuming processes. They must know everything from required capabilities to schedules and development status and interoperability standards (including data elements) employed. They must have access to analyses of performance in various mission contexts.

DOD should develop minimum posting requirements for all system developments. These might include system requirements, schedules, interoperability standards employed, and program status. Individual SOSEs might want to augment this list for the SOSs under their purview. To keep posted information timely (thus further reducing developmental friction), DOD should deploy developmental tools that automatically post results. For example, proposed schedule slips or proposed performance changes should be posted as they are proposed, automatically, with the proposal status attached, so that other system developers are not surprised when changes occur. Posting information as it is developed and while it is awaiting approval is a radical cultural change that will require high-level support on a continuing basis—but it will greatly improve cross-program coordination and reduce surprise caused by schedule slippage.

In addition, interoperability and timely systems development are often hampered by lack of knowledge (e.g., of the interoperability standards employed) of the systems with which a system under development must interact. Operational, technical, and systems architectures for DOD should be developed with inputs from across DOD. The systems architectures should be linked to the Web sites of the individual systems, again to improve visibility across the department. The technical architecture needs to focus on standards for Web services to enable a truly joint, net-centric enterprise service architecture to evolve.

System-of-systems engineers need to perform cost-performance-agility trade-offs across systems, and individual systems developers need to understand what features of their systems improve mission capabilities. This requires contextual performance modeling capabilities (models that relate system performance, in the context of other systems, to some higher level of mission accomplishment) that are shared within communities of interest, and occasionally across them. Thus, modeling and simulation across systems-of-systems must be made easy. This is best accomplished if both sponsoring organizations and systems engineers agree on the modeling framework, shared data, and performance measures, and if they sponsor tools that will enable individual systems to develop models that can fit into the framework and be used by others. The Joint Staff NETWARS C3 model has already used this approach successfully to provide an integrated communications and applications model for DOD C3.

To develop systems (and net-centric services) that are interoperable, developers must have access to contextual design and development environments that include current systems and the developmental versions (hardware, software, and computer models) of future systems with which they must interoperate. In addition, the challenge of development is frequently hindered by interoperability problems that are not discovered until testing. Because of the large number of systems under simultaneous development, interoperability and networked capability development can only be accomplished by the use of distributed modeling, development, and test environments. However, its implementation depends on significant funding, a dedicated effort to resolve security and clearance problems, and potentially an operational mandate by the Joint Staff.

DOD interoperability would be greatly enhanced by the use of profiles of commercial interoperability standards and commercial interoperability processes. However, unless the organizations that must abide by these standards have significant input into their selection and development (an important cultural change), many developers will devote significant effort to raising issues over whether they should be bound to standards that reduce their individual effectiveness, and DOD will devote significant effort to the issue resolution process. Better to broaden DOD's involvement in the development and adoption of standards and provide specific tools (e.g., common test facilities and statements that can be added to requests for proposals to ensure compliance) to make following the guidance easier.

CULTURE

The new culture must be one of common purpose and information sharing. It should proceed from the notion that all systems must perform in the context of other systems, so that the capabilities they produce are dependent on how well these systems work together. Thus, at every level of the three major DOD processes, but especially in the systems development process, reviews must begin with questions of performance in context. This places an emphasis on systems analysis (rather than programmatic cost and schedule) and requires that SOSAs have SOSEs who can address systems questions in the broader context.

For these SOSAs and SOSEs to answer such contextual mission and performance-oriented questions, DOD will have to adopt an information-sharing culture. It will be the norm to post and share information and the norm to work joint modeling efforts with the other systems with which a system must interoperate in performance of a mission. This new culture must be driven from the top and requires both rewards for sharing information and penalties for not being forthcoming.

A potential drawback of fully networking the members of the developmental processes of course occurs if they are competing rather than collaborating. Such competition can only be prevented by the attention of senior leadership at the highest levels to the creation of unity of purpose.

Most fundamentally, at all levels from senior policymakers on down, SOSAs should expect to be asked contextual, capabilities-based questions and should automatically create SOSEs to help them answer those questions. These SOSEs should know their three roles, be able to find and collaborate with the SOSEs of related systems, and expect that the SOSEs they work with are similarly trained and prepared.

ENTERPRISE-WIDE SYSTEM-OF-SYSTEMS ENGINEERING ORGANIZATION

All this will not happen spontaneously or by fiat. DOD, and especially its Global Information Grid, will need a focal point activity—a high-level system-of-systems engineer whose focus is not on doing systems engineering and setting design criteria, but rather on creating the culture and collaborative systems engineering environments that will enable DOD systems engineering.

This enterprise-wide system-of-systems engineering organization should, with sponsorship of the ASD/NII and the USD/AT&L (undersecretary of defense for acquisition, technology, and logistics), create and lead a council of SOSAs and SOSEs that can define many of the specific portions of the collaborative environment. For example, this council can set the posting requirements, define what a DOD interoperability standards program should look like, define the desired net-centric enterprise services environment, and unify its members' programs.

Some key capabilities (e.g., communications) do not yet have defined systems-of-systems and SOSEs. For such key capabilities, the council might appoint champions to facilitate the meeting of appropriate SOSEs and systems engineers, so that analyses can proceed across these systems.

This organization could also reenergize DOD's standards and systems integration processes, with participation across DOD, with an emphasis on simpler processes and common agreements. It could help energize and unify performance modeling activities and contextual design, development, and test activities across DOD.

CONCLUSION

The development of better enterprise-wide processes in DOD for simultaneous systems development in an increasingly fast-paced and complex world has been hampered by the lack of a conceptual framework and vocabulary for dealing with the major underlying challenges. This chapter provides such a conceptual framework. It analyzes the underlying problems, proposes guiding principles that are an extension of net-centricity, and makes specific recommendations to ameliorate the major system-of-systems development challenges.

These proposals will ultimately reduce developmental friction, improve decisionmaking, and unleash the creativity of the "edge" of the broader DOD development community—that is, every person working on every aspect of every problem—by enabling them to take initiatives faster while in the context of overall (but evolving) goals, without the limitations imposed by stove-piped information flows.

NOTES

1. The views expressed in this chapter are those of the author and do not reflect the official policy or position of the National Defense University, the Defense Information Systems Agency, the Department of Defense, or the U.S. government. All information and sources for this chapter were drawn from unclassified materials.

2. See Jeremy Kaplan, *A New Conceptual Framework for Net-Centric, Enterprise-Wide System-of-Systems Engineering,* National Defense University, Defense and Technology Paper no. 30, at http://www.ndu.edu/ctnsp/Defense_Tech_Papers.htm.

ENGINEERING OF COMPLEX SYSTEMS
CHALLENGES IN THE THEORY AND PRACTICE

DOUGLAS O. NORMAN

Some view systems engineering as consisting of technical aspects only. Often it is argued that optimizing or replacing particular technologies (or particular technical processes) will enable us to scale up systems engineering to enterprise-sized entities. This ignores the assessment of value—where, when, why, and by whom—that directly affects the definition of engineering processes and activities and modulates their application.

This chapter argues that the nature of large systems and the manner and place of value assessment conspire to diminish the effectiveness of the current acquisition system, especially as it is applied to information technology (IT)–intensive systems. The acquisition system sets the engineering goals and tasks and, in doing so within the current processes, sows the seeds of failure. There must be changes that acknowledge some fundamental aspects in the environment in which the acquisitions are intended to provide systems to regain effectiveness. Our language demonstrates that we understand this. We do not usually refer to "systems"—we refer to "solutions." The word "solutions" suggests that we do not equate a system with a solution to a problem, yet the insistence on equating the two—even implicitly—renders the judgment that there is one solution to a problem, applicable to all who share the problem.

The argument starts with (what here will be) an assertion, which has been well-argued elsewhere—that IT-intensive systems, such as command and control systems in the DOD, behave as complex adaptive systems. This is largely due to the broad coverage of use and users—both vertically in organizations and horizontally across organizations—and the changing nature of the users' goals and priorities while employing the system(s). The nature of the changing environment also impinges on the system needs and requirements. These characteristics demand change and adaptation in those systems applied to environments that themselves undergo continuous change and adaptation. For example, the DOD is itself a *complex adaptive system*, and an aspect of it (such as C2) that attempts to bridge the whole of it must also be a *complex adaptive system*.

A key aspect of any complex system is that its points of stability are of the dynamic, not static, variety. Thus any change in the underlying "tension" among all the stakeholders or the environment causes a movement in the point of dynamic stability. Often, these points and places cannot be predicted, nor can the timing of the introduction of change. Consider that many times stakeholders are not within local communities; they are separate, and even perhaps uncommunicative, yet they get a "vote" and can introduce changes asynchronously and unwanted by some. The continuous, unpredictable changes cause difficulties in many things, not least in articulating the *value* resulting from pursuing any particular materiel acquisition.

Value, and its assessment, drives acquisition processes and decisions. What will be argued here is that the current processes and the proxies for value that are used have hijacked the ability to focus on demonstrated operational value in a timely fashion, which is not only the initial, but also the ultimate, goal.

Acquisitions are undertaken to provide materiel solutions to operational entities so that they may accomplish the missions with which they are tasked. Thus, the most important aspect of "value" is when it is realized in an operational setting, not when specifications for a notional thing of operational value are verified as being present. It is too likely that a disconnect exists between what is needed (and therefore of value) and what the formal requirements deliver (which is not guaranteed to cohere to the real need).

There are many proxies for this *operational value* that are used to partition the acquisition process into manageable chunks that run essentially independently of one another. Unfortunately, the partitioning has resulted in a diminution of the primary value sought— demonstrated operational value. This diminution occurs because of the transformations made as an operational need is translated into a set of requirements and as the set of requirements is translated into a design that is then contracted for construction. As in the child's game of telephone, changes (i.e., mistranslations) are introduced along the chain of transformations.

Another set of characteristics that militates against success involves the expense of defining, contracting, building, and delivering the complete materiel solution. The resulting attempt to ensure that we have all the requirements (i.e., *complete*) has the effect of a positive-feedback loop, further increasing time and expense and encouraging the introduction of additional requirements. (The thinking is that if we miss the opportunity, when will we get another chance? After all, we get only one bite of the apple.) This requirements pile-on effect is seen quite often, and it drives up the likely cost and time. An additional outcome of the bloat in requirements, time, and expense is the involvement of additional levels of bureaucracy. We are spending public dollars, and the taxpayers have every right to expect that there is good justification for their money to be spent in a particular way. Thus, more oversight is provided for such assurances. Not surprisingly, this too extends time and expense.

Materiel solutions do not just happen; they are engineered. Because the larger process within which the engineering is performed offers proxies to the engineers, as contrasted to the initial or ultimate goals, the process used by the engineers supports optimizing the value for the proxy. As already shown, this involves a fixed set of requirements (often large) against which a design is created. Because the set of requirements is presumed to be a valid transformation of needs, there is clear desire for optimizing cost, schedule, and performance against these preexisting, stable requirements.

At this point, the discontinuity shows itself vividly. The engineering processes assume and require a point of stability around the requirements, and the requirements are assumed and required to be a stable transformation around needs. But these value proxies and goals do not form a stable relationship in the operational world that is subject to continuous change—rather they recall only a point of dynamic stability (real or imagined) that generated the initial statement of operational need.

It appears that the problem is overconstrained and has no answer. Spending public money requires a clear goal for satisfying an operational need and an executable plan to achieve it. Yet the time and expense to ensure that one has such things for all but the trivial seems unachievable before changes in the field render our attempt "overcome by events."

Now the real questions stand ready: (1) how might one perform engineering and acquisition *within a complex environment*, and (2) how might one perform engineering and acquisition *of a complex system*?

We should review what it means to be *within a complex environment*. If multiple external elements (people, organizations, technological artifacts not under local control) impact or influence your ability to establish and enforce success criteria, system characteristics, value measures, etc., then you are operating *within a complex environment*.

So, if operating *within a complex environment*, how does one mitigate the perturbations likely introduced? Two main strategies are used today: (1) attempt to ignore the external influences, or (2) seek consensus among all the stakeholders. Both approaches tend to produce less-than-stellar results.

The first strategy, an isolationist approach, assumes that the proxy for operational value (i.e., the engineering requirements from which all the planning flowed—including cost and schedule estimates) is sufficient for delivering operational value. The attempt here is to provide a stable environment within the system development activities by protecting them from external "noise." In essence: ignore the enterprise. This can result in individual program "success" but enterprise failure.

The second strategy is a minimax play across all stakeholder value functions. The issue with this strategy is that the minimax optimality point assumes stable value functions, which is seldom the case. A substrategy often introduced to diminish this possibility is to issue a memorandum of agreement (MOA) or a memorandum of understanding (MOU) that outlines the various stakeholder organizations' responsibilities and defines the processes and protocols for introducing and managing change. Although superficially a good thing, the results seem to end up as a bloated, turgid process for slowing down the introduction of change, thus attempting to provide the stable environment (see the first strategy) to the development stakeholders. This seems to result from the nature of the interactions among the stakeholder groups; they are engaging in defensive moves and protecting their equities—lowering the possibility that they might receive blame if things go wrong and ensuring that they get the most return on any investment they have made. This is not an unreasonable response, given that so often systems fielded at the end of a long acquisition cycle are effectively rejected once they hit the field. Another substrategy is for a politically powerful stakeholder to wield a big stick and attempt to force consensus.

Because in both strategies the external influences are applied to the whole of the system under development, the result of applying either strategy applies to the whole system. Thus everything succeeds or fails as a whole. Generally, either of the two strategies moves toward failure. The first increases the likelihood that we will deliver a solution in the future for a need no longer present. The second increases the likelihood of development stasis caused by stakeholder changes of mind, resulting in requirements holes or requirements invalidations and a whipsawing of development activities (recall that stable points among independent stakeholders are dynamic, not static).

These behaviors may result because of the weaknesses of the operational value proxies. Assuming this, the answer may be actually quite simple: let the operational community choose. It is their understanding of demonstrated operational value that systems attempt to satisfy, and it should be their sense of value that should be applied. This may appear at first blush to be dangerous as it would seem to give the operational user the ability to reject that for which much money, time, and resources have been expended. The better question might be: how do we change the manner in which we build things such that we can apply operational value assessments early and often? The answer to this starts to answer the second major question: how might one perform engineering and acquisition *of a complex system*?

The strategy staring us in the face is to "focus on demonstrated, realized *operational value*." From an engineering point of view, this assessment, which occurs post-development, would seem to raise risk. This is true, but only if we make our bets on the level of the whole system, which sadly often happens today anyway. Suppose we bet on something smaller. Rather than focusing on a whole-system basis and in isolation from the enterprise, suppose we focused on the potential unique value our particular key delivery to the enterprise offered. Suppose we focused on the nature of how our key unique value contribution composed with other elements of value to allow and render new capabilities.

For its part, the "enterprise" must come to reward those who deliver valued elements (those offering demonstrated operational value as determined by operational use and assessment). This strategy implies a value stream (likely money) flowing to those who seem to be able to understand need and deliver technical elements that are chosen by those performing an operational role. It suggests choice, not singletons, as potential technical elements, and it also suggests elimination of those elements that are not chosen—i.e., programmatic death rather than defense and support. It all starts to sound like a "marketplace." But what about the taxpayer and the demand to pay only for that which has value? Marketplaces have many overlapping offerings from which buyers choose. Is this not wasteful? It certainly would be too expensive to pay to build multiple solutions to each given problem. This suggests that we do not (perhaps) pay for potential solutions with public money, just for those that have demonstrated value. Thus payment plans reward the valued, delivered goods; and there would be no delivered goods if they did not interoperate with elements already in the field.

How should we structure payment plans to reward realized accomplishment (value) in the operational domain? How should we shift from "cost" to "price"? How do we re-jigger the incentive structures? These are the questions for our consideration as we look to reinvigorate systems engineering and recognize the breadth of its influence.

Providing useful answers to these questions is not only possible, it is eminently doable—today.

MANAGING MEGAPROJECTS
LESSONS FOR FUTURE COMBAT SYSTEMS

MARCO IANSITI

Future Combat Systems (FCS), the cornerstone of the Army's "Future Force" modernization program,[1] is among the most complex, uncertain, and distributed projects ever attempted in human history. This chapter considers the ability of the Lead Systems Integrator (LSI) to navigate the complex, uncertain, and distributed requirements of FCS through management practices that ensure successful performance in each of these capability dimensions.

Because FCS shares many of the most important characteristics of megaprojects, the management of FCS projects should be informed by an understanding of the management strategies used in megaprojects that achieve to successful outcomes.[2]

To that end, we studied the management practices used in 31 technology-intensive megaprojects, screened to ensure that the challenges are similar to those of the FCS. For these megaprojects, the study evaluated the relationship between the program's management practices and its outcomes to determine which management practices were successful. The study's results show a very high level of correlation between overall management capabilities and overall program outcome.[3] They also show a significant difference in management performance between those companies that had successful outcomes and those that did not. In fact, the programs in the top quartile of project outcome exhibited specific management capabilities that clearly differentiated them from programs in the bottom quartile. Finally, a level deeper, each of the primary management concepts considered in this study—program management, ecosystem management, and process integrity—also show a high degree of correlation with overall program outcome.

A megaproject also encounters, and must overcome, the massive technological and systemic changes in the innovation environment. Program management still requires the consistent and disciplined coordination that has always been its touchstone. But now, so many organizations are linked, and the interaction of these firms is a complex phenomenon that requires the cooperation and motivation of a vast network of interrelated products, services, and technologies. In this way, both the responsibilities and the outcomes of customers, suppliers, partners, competitors, and many others must be managed in an expert fashion with a true understanding of their interrelation in order to bring about the desired outcome.

In sum, an LSI provides FCS with the managerial capabilities that have been critical to achieving successful outcomes in similar megaprojects.

MEGAPROJECTS AND FCS FACE SIMILAR CHALLENGES AND RISKS OF FAILURE

As the great disparity among competitors evaluated in the study suggests, many megaprojects failed to meet their performance obligations, and such failures caused significant downward departures in outcomes. Table 8.1 provides some examples of major failures in recent, high-profile megaprojects.[4]

GENERAL MEGAPROJECT CHALLENGES

Failures like those described above are borne of the unique challenges faced by megaprojects. As our study indicates, those challenges are best conquered through the skillful execution of each of three fundamental management frameworks discussed below. Although a project of any size or complexity is subject to its own set of challenges, a megaproject such as FCS faces a myriad of individual and collective obstacles to a degree that can seriously jeopardize the ability of that project to meet expectations. Specifically, a megaproject will involve substantial risks of failure as result of each of these three categories:

- *Complexity,* or the program's sheer size, duration, and scale. Indicators include overall budget, overall schedule, the number

Table 8.1. Performance Outcomes in Selected Contemporary Megaprojects

Project	Cost Overrun (percent)	Schedule	Performance
Channel tunnel	80	2 years late	Half of projected revenues
Denver airport	300	16 months late	Half of projected traffic
Big Dig	250	5 years late (and counting?)	Still leaking, unfinished
Iridium	71	2 years late	Less than one-tenth projected users in first year
Average for large-scale projects	88	17 percent schedule slippage	72 percent fail to reach profit objectives

Source: See note 4 at the end of this chapter.

of companies and people involved in the project, the use of new or unproven technology, the number of systems or elements that need to be integrated, and the number of lines of code used in the project.

- *Uncertainty,* or the external forces and *events* that can influence the direction, execution, or outcome of the program. These can include the insertion or changing of technologies in the program (based on external technology evolution), externally imposed changes to program budget, schedule, or scope (for example, by Congress, the executive branch, or other government agencies), emerging market requirements, revealed and evolving enemy threats, or changing world events (for example, the fall of the Berlin Wall or the space shuttle *Challenger* disaster). Uncertainties are realized over time as programs evolve, and in some cases (as with FCS), uncertainty can actually increase over time.

- *Distributed nature,* or the large number of separate organizations and divisions acting as stakeholders in the program. These organizations may be distributed geographically, but more crucially decisionmaking and execution authority are shared among many organizations in programs of this type.

Managing Megaprojects 91

Figure 8.1. Megaprojects Challenge Dimensions

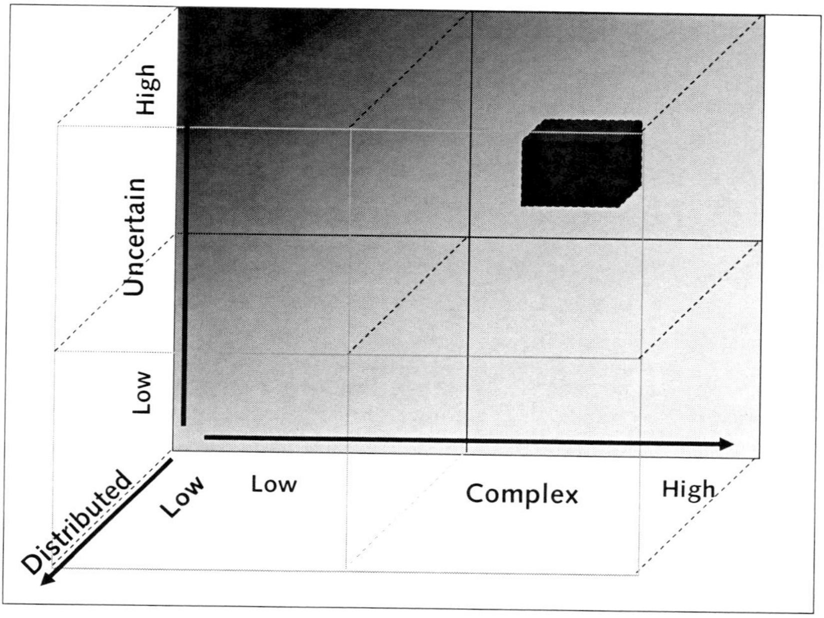

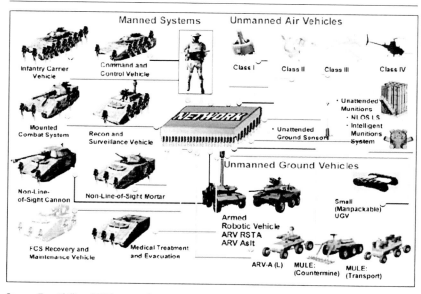

Figure 8.2. FCS Complexity

Source: Tom Phillips, *FCS Program Overview,* PowerPoint presentation, March 11, 2005, http://www.boeing.com/defense-space/ic/fcs/bia/050311_ovrview_forum.pdf.

Figure 8.1 maps projects based on the existence and magnitude of the challenges described above. A megaproject, represented by the dark box in the upper right quadrant, has the greatest risk of failure because it combines high uncertainty, high complexity, and a highly distributed nature.

Our study considers and assigns a value that measures the magnitude of each challenge dimension for a given megaproject. However, given that the megaprojects studied all score high in each challenge dimension, it can be said that they all operate in an environment where significant risks of failure are present.

FCS CHALLENGES

This section reviews and graphically depicts the challenges facing FCS with respect to its complexity, uncertainty, and distributed nature. FCS faces significant challenges in each dimension, and those challenges are similar in nature to those in the megaprojects included in our study. In fact, the challenges presented by FCS tended to be higher than those typically experienced in a megaproject. Therefore, it may be even more important that FCS achieve high levels of performance in each of the managerial capability dimensions that proved so important to project success in our study.

FCS has a high degree of complexity in its number of participants, but, even more important, in its technology requirements. There are currently 21 companies with design and build contracts awarded, and approximately 360 organizations are involved overall.[5] The FCS technology requirements are even more daunting than the sheer volume of the program's organization. Many of the FCS component technologies are highly sophisticated and evolving. Technologies will be constantly matured and incorporated over time and, undoubtedly, technologies that have not even been conceived yet will be in the final products. This sophistication exponentially complicates the FCS's interoperability requirements. An estimated 80 million lines of code will be required to integrate the complex systems (18+network+soldier), twice the number of lines of code in Windows XP.[6] And the consistent performance of this integrated code will be necessary if FCS is to achieve a successful outcome.

Figure 8.2 depicts one element of FCS complexity by showing the diverse and complex network management requirements associated with manned systems and unmanned air and ground vehicles.

Figure 8.3. FCS Uncertainty

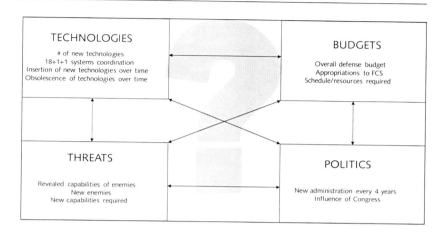

FCS also has significant uncertainty in each of its core elements: technology, budget, threats addressed, and politics. Furthermore, each of these factors is dynamic and interconnected, multiplying the effect of this uncertainty and presenting significant risks of failure. Figure 8.3 details several important factors creating uncertainty in just four of the FCS's elements.

FCS is also one of the most distributed projects ever attempted by the military or elsewhere. To succeed, it will require the concerted effort of stakeholders in industry and an array of government authorities, all with competing agendas and divergent goals. FCS's geographical diversity, depicted in figure 8.4, is but one measure of its disparate nature and attendant management challenges.[7]

Taken together, as figure 8.5 shows, the challenges to FCS place it within the project quadrant exhibiting the highest risk of failure.

MANAGERIAL CAPABILITIES CRITICAL TO SUCCESSFUL OUTCOMES

The ability to meet or overcome the significant challenges facing megaprojects and therefore achieve a successful outcome is closely correlated with that megaproject's management capabilities. Our study relied upon a number of theoretical frameworks in the areas of program management, business ecosystems, innovation, and product development to develop the appropriate set of management capabilities by which management performance was to be determined.

Figure 8.4. FCS's Distributed Nature

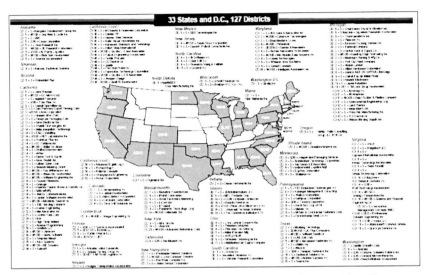

Source: Phillips, *FCS Program Overview*, PowerPoint presentation, March 11, 2005, http://www.boeing.com/defense-space/ic/fcs/bia/050311_ovrview_forum.pdf.

Figure 8.5. FCS's Risks of Failure (Combined View)

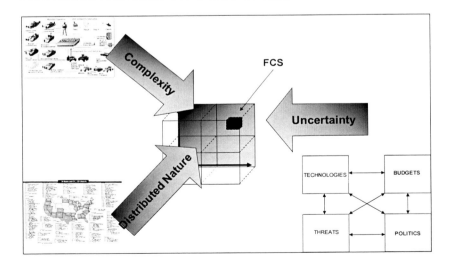

Program management theory, insofar as it pertains to management within the walls of a single organization, is well established.[8] In executing these strategies, four modes of program management structure are common: functional, lightweight, premier, and autonomous.[9] The complexity of megaprojects requires premier program management, as it requires advanced execution in aspects of both of these strategies in order to accomplish successful program outcomes.

Figure 8.6 displays the evolution of project management requirements as the complexity, uncertainty, and distributed nature of a given project grows. When each of these factors is high, a megaproject likely requires the premier program management structure depicted on the far right.

A component of such premier program management is the strong control and influence of an effective project management team. But, as discussed below, premier program management is only one of three management frameworks critical to the success of megaprojects.

Ecosystem management can be thought of as program management beyond the boundaries of one's own organization. This study draws heavily on previous and ongoing research in this area.[10] Managing a large, highly distributed network of partners, suppliers, customers, and other stakeholders is a crucial but challenging undertaking. A "keystone" organization serves as a hub for the other ecosystem members. By regulating the connections within the network and maintaining a predictable platform on which other network members can rely, keystones have an outsized effect on the performance of the entire system. Keystones have been shown to be critical factors in boosting the overall productivity of a network, increasing its robustness in the face of outside threats, and promoting innovation. By improving the health of the ecosystem in which they reside, keystones sustain their own existence. On the other hand, poorly functioning ecosystem hubs can reduce the health of their ecosystems and even imperil the survival of themselves and every member of their network. When a keystone is suddenly removed from an ecosystem, a dramatic and chaotic reorganization ensues.

Figure 8.7 shows how a keystone combines strategies related to platforms, standards, and common architecture to assist, guide, and ultimately maximize the performance of the ecosystem players with which it interacts.

Process integrity, or *process design*, refers to strategies used to manage, retain, and apply knowledge across the phases of the project (such

Figure 8.6. Organizational Structure Matrix

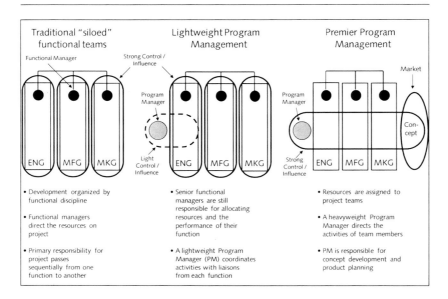

Figure 8.7. Properties of Business Ecosystems

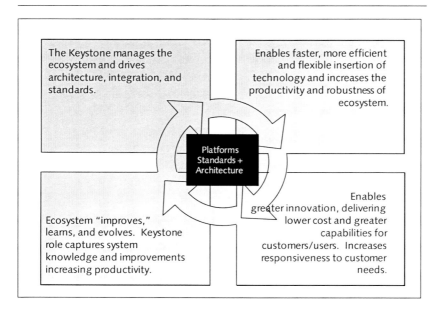

Figure 8.8. Process Design

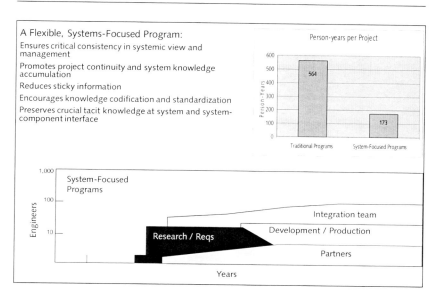

as research and development, pre-production, production, operations, sustainment, and others, depending on the type of project). Megaprojects face significantly greater complexity and uncertainty in phase transitions than more traditional programs because of their long durations and frequent overlap of design, production, and fielding. The author has studied and written on the challenges faced in "traditional" high-technology development projects, such as mainframe processor modules, semiconductors, and workstations, and the technology integration strategies used to address them.[11] For this study, we adapt and apply those frameworks to analyze how megaprojects utilize systems integration knowledge, prototyping, experimentation, and simulation across phase transitions.

Figure 8.8 details the numerous advantages of a flexible systems-focused program management strategy. Figure 8.8 also shows through a comparison of person-years per project to traditional programs and a timeline perspective that such a flexible, systems-focused approach confers these benefits more efficiently without sacrificing the consistency essential to a megaproject's success.

Figure 8.9 summarizes the theoretical frameworks, or managerial levels, from which this study derived the managerial capabilities that measure performance in each area as well as overall.

Figure 8.9. Critical Managerial Theoretical Frameworks

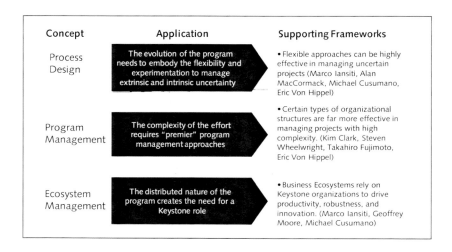

The next section reviews this study's specific findings with respect to megaprojects performance along each of these managerial concepts and relates them to the LSI's role in achieving similar managerial objectives in FCS.

LSI OPPORTUNITIES FOR FCS

Given the high degree of uncertainty, complexity, and its distributed nature, the FCS, like the megaprojects in this study, requires successful execution in three dimensions—premier program management, ecosystem management, and process integrity. Our study developed specific categories, or subdimensions, of management capabilities to measure a given megaproject's execution ability in each of the three main theoretical dimensions. As figure 8.10 shows, those megaprojects in the top-performing quartile showed consistently superior management capabilities.

This section reviews specific examples of the management capabilities identified in the study as necessary to a megaproject's execution in each of the key theoretical frameworks. It also discusses the roles and responsibilities of successful LSIs and shows that they can provide the technical expertise, program management, and systems engineering experience required to help FCS successfully exhibit these critical capabilities.

Figure 8.10. Megaproject Performance on Management Capabilities

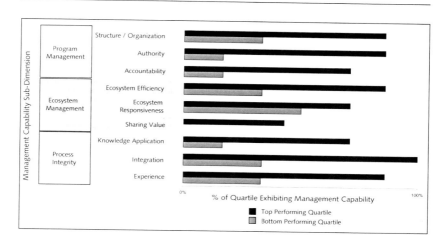

PREMIER PROGRAM MANAGEMENT

Structure and Organization

Our study indicates that successful megaprojects exhibited a structure that has evolved from mere "coordination" to a "fully integrated" approach, especially given the complexity of megaprojects.[12] This is consistent with the conclusions in numerous research sources. Although the elements of a coordinating program management organization essentially fulfilling an administrative role in managing program schedule, budget, contracts, and general organization of various project groups and teams are necessary for a successful program, they are no longer sufficient. A fully integrated organization can provide these coordination functions while it maintains a horizontal, integrative viewpoint, actively engages with project groups and teams, and plays a significant role in problem solving and decisionmaking. We even observed that the evolution from a coordination organization to a fully integrated one can sometimes happen over the course of a project. Some megaprojects were able to withstand these growing pains, but it is surely better to adopt the proper organizational structure with the proper personnel from the outset of a project.

In our study, it became clear that although many megaprojects might strive to provide a fully integrated organization, those that succeed have a project management team with deep systemic knowledge that reaches across the most critical areas of the project as well as the ability to communicate that vision and knowledge.[13] Our study's findings also confirm that the way in which multiple offices and firms and their relationships are structured is a critical aspect of program success. A fully integrated organization structure should have the ability to accommodate highly distributed projects. Organizational barriers are typically broken down to the extent possible to align all players toward a common goal. The customer organization is closely involved with a single, fully integrated, authoritative, accountable program management team (such as in FCS) which has decisionmaking authority and responsibility for interfacing with the customer.

The LSI provides FCS with fully integrated program management through its deep domain expertise. The LSI identifies and solves problems in individual projects, of course. But rather than provide such solutions in a discrete fashion, its sophisticated understanding of FCS ensures that those solutions successfully integrate with the project's goals and direction. This goes beyond pulling in domain experts on an intermittent basis when needed. The LSI also fulfills its role as a lead communicator of this vision throughout the FCS project. Its consistent leadership helps instill a culture that values the performance of the overall system over the individual components, which helps lead to better discovery of innovations and more alignment on standards across the ecosystem. This communication role is critical to fully integrated project management. Finally, FCS requires a fully integrated organizational approach because it not only manages a plethora of outside firms but multiple governmental program management offices.[14] Combine that with the management of the various government entities at the local, state, and national level, and again one sees that FCS is even more complex and challenging than most megaprojects. This complexity is simply beyond the ambit of mere "coordination," especially when one considers the additional layers of management capabilities exhibited by an LSI, as outlined below.

Authority

Our study found that top-performing megaprojects were much more likely to have a program management organization with significant authority. Our study adopted the generally accepted concept of program management authority to measure the degree of authority given to the team in each megaproject. Program management authority considers the following factors: responsibility for making decisions on specifications, responsibility for implementing those specifications in the product, and the ability to direct and control a project's resources. The study revealed that authority is driven by work content and reputation rather than the program manager's position in the organization. Thus, authority is somewhat dynamic and will expand or contract based on management's performance.

In many cases, particularly with government-funded and directed programs, the group defining requirements is distinct from the group actually executing and building the product. In such cases, the program can still be scored positively on this subdimension, but there must be a rich flow of information and strong alignment of objectives between the two groups. This requires deep and constant communication and interaction between the two groups, an effective working relationship, and alignment on project goals and technical capabilities.

FCS has more program management organizations than any other project in the study and therefore is in the most danger of leaving its program management team without authority. As the study finds, such lack of authority can lead to a megaproject's failure to meet its objectives. To maintain authority even in this disparate environment, Boeing (as FSC LSI) implemented a OneTeam model that colocates the program management teams of the government and LSI, actually embedding government personnel in the integrated product teams. This model has helped ensure that requirement-setting, an essential component of authority, is accomplished in an informed and collaborative way across the Army program office, the LSI, and the contractors responsible for the development of FCS systems.

Accountability

Megaprojects led by program management teams accountable through all execution stages up to and including quality assurance and testing tended to have more positive program outcomes. Premier project management also ensures that outcomes of each such execution stage are consistent with the desired final outcome.

The Army has firmly established that the LSI is accountable for final project outcomes, including quality assurance and testing of the FCS. Therefore, the LSI needs to stay accountable for later phases of the project to drive a positive outcome.[15]

ECOSYSTEM MANAGEMENT

Ecosystem Efficiency

To be successful, modern program management must recognize and exploit the complex network, or ecosystem, that both directly and indirectly impacts a project's outcomes. By regulating the connections within the network and maintaining a predictable platform on which other network members can rely, keystone organizations can achieve an outsized and positive effect on the performance of the entire system. Keystones have been shown to be critical factors in boosting the overall productivity of a network, increasing its robustness in the face of outside threats, and promoting innovation. An ecosystem left to its own devices, on the other hand, will not only cause a megaproject to miss out on enormous opportunities, but may actually seriously undermine a project and lead to its ultimate failure. In an effort to define those attributes that define successful keystone project management, our study identifies and measures five primary strategies that can increase the efficiency of ecosystems: (1) coordinating communication, (2) providing tools for innovation, (3) providing financial leverage, (4) communicating customer requirements, and (5) sharing assets.

Megaprojects, which have distributed and complex business ecosystems, require frequent, open channels of communication that foster and coordinate communication in the form of the exchange of critical information. Successful program management takes an aggressive posture to maximize its information efficiencies. For example, our study found megaprojects that convened special teams with representatives from key organizations with the explicit task of ensuring and facilitating understanding of interdependencies and interconnects between elements. Others set up advanced video teleconferencing systems and have used them extensively (virtually around the clock) to tie together teams from disparate locations.

Successful program management can actively enable innovation in a given project's ecosystem. Many programs share software design tools that were developed internally by a prime contractor. These tools may provide functionality not available in commercial products and can improve the innovations made by subcontractors. This can significantly increase the productivity of the entire system and help lead to successful project outcomes.

The clear definition and communication of client requirements greatly assists the actual meeting of those requirements. A program management team can play an important role in ensuring that both the client and service organizations support and work to achieve the same requirements. First, the program management team can work with the client to ensure that the requirements fit their needs and are achievable given the scope of the program. Second, that same team can utilize their understanding and first-hand knowledge to effectively communicate these requirements to the design, development, and manufacturing organizations charged with delivering the products to meet those requirements.[16]

Finally, keystone organizations recognize that more can be gained by sharing, rather than hoarding, its assets. It is a common practice in federal programs, for example, for the government to provide testing facilities for contractors to use, which clearly drives efficiencies by significant costs of product development while increasing standardization and the use of common platforms that are so critical to eventual project success. On a smaller scale, successful project management teams will often make their expert personnel available to ecosystem members. This practice increases the impact of their work and averts potential problems by identifying them in early stages before they become costly and less manageable.

Faced with perhaps the largest, most diverse, and most complex ecosystem of any of the megaprojects in this study, the LSI has made numerous investments in ecosystem management. For example, it created and matured its OneTeam structure, which employs a number of these strategies to improve the efficiency of FCS's ecosystem. For example, the LSI developed and made available APS technology, which continues to spur ecosystem innovation that will significantly enhance the FCS program outcomes.[17]

Ecosystem Responsiveness

The program management teams at top-performing megaprojects used strategies that *aligned* the innovation, efficiency, and productivity of their ecosystem organizations toward shared goals rather than individual interests. Strategies include (1) the use of information hubs, (2) creation and enforcement of integration standards, and (3) tracking the performance of ecosystem members. This section reviews some of the ways that Boeing, as FCS LSI, took advantage of these strategies to the benefit of the FCS project.

In its role as FCS LSI, Boeing's Advanced Collaborative Environment (ACE) builds upon similar hubs used successfully in other megaprojects to support ecosystem responsiveness. ACE's requirements help firms manage fluctuations in program standards and respond to changes in the ecosystem, while still maintaining systems-level consistency. This ensures that organizations are working toward systems-level benefits and performance rather than individual self-interests.

The development and deployment of proper standards across an ecosystem can produce a more serviceable and affordable end product for the customer. The FCS's decision to use line replaceable modules (LRMs) rather than line replaceable units (LRUs) is just such an example. Both Boeing (as FCS LSI) and the Army should be applauded for its decision with respect to LRMs, particularly because though they are more costly at the inception, they will save approximately $4 billion in the operations and support phase of the project.[18] These long-term savings are more often achieved when the program management team's incentives are aligned with the long-term outcomes desired by the client, as they are in the case of a proper LSI who will see a project through completion.

Finally, the LSI has taken significant steps to measure and enforce performance standards. The CEO Council, a subcommittee of the OneTeam, consists of the CEOs of all involved organizations. The CEO Council meets once every four months and reviews the performance of each company. Enabled by detailed tracking of performance on multiple dimensions for each company, public display of a "red" rating in front of a CEO's peers is a powerful mechanism for driving performance in contracting organizations. This is an excellent example of measuring and driving high performance across an ecosystem.

PROCESS INTEGRITY

Integration

The megaproject leaders that participated in our study consistently identified process integration as one of the most important management capabilities. Top-performing organizations tackle integration by emphasizing informed decisionmaking. They tend to place responsibility for critical decisions in a group with the relevant systems and integration knowledge. The importance of a process integrity group and of robust experimentation increases as a program changes over time, which is when you see the greatest overlap between design, development, manufacturing, fielding, and sustainment.

The integration necessary to maintain process integrity is a complex problem requiring multifaceted solutions. Our study uncovered many successful efforts to solve this difficult problem. In some cases, systems engineering organizations were given explicit responsibility for integration, with the design and manufacturing organizations simultaneously reporting to a single engineering chief. At least one company we spoke with has established a corporate-wide knowledge management group to develop strategies and techniques to handle integration tasks across all programs.

The consistency that an LSI brings to a project provides, when coupled with a commitment to flexibility in the face of technological and other changes, an antidote to the complexity and uncertainty that makes process integrity so difficult.

The FCS LSI has established both production and support planning teams with budgets and mandates to handle transition and integration issues. While some challenges exist because no contractual tool is yet in place for production, the LSI has architected a program that would enable effective transition to management strategy upon FCS's completion.

Experience and Experimentation
Attracting, utilizing, growing, and retaining individuals with experience in similar systems integration programs is critical to the process integrity and ultimate success of large-scale programs. Certainly every program claims to try to hire the best and the brightest, but our study found that more successful megaprojects invested more time and resources in developing and maintaining more robust hiring and retention programs. Interestingly, a virtuous cycle exists in successful programs that manage sticky information by leveraging experimentation extensively and frequently throughout the project. These programs retain talented individuals because they encourage innovation and creativity. These individuals then perform better *and* stay longer, increasing the program team's institutional knowledge, which again improves performance.

Although government organizations may not be able to match private industry financial compensation, they certainly can excel in the other very important strategies that lead to better recruitment and retention. For example, a firm created a "school" for all new recruits that provided several weeks of program management training, which led to employees with a broader view of how complex, multisystem problems were managed. This broader, more holistic understanding is critical to the fully integrated organizational structure discussed above.

Of course, experimentation alone is critical to problem solving and can lead to breakthrough innovations, both of which tend to lead to successful program outcomes. By devoting as much capacity as possible to simulating the entire system, integration groups are able to test multiple configurations before having to commit to a final design. This supports responsiveness to market needs and drives higher levels of system quality. Flexible development processes that incorporate continuous experimentation and rapid prototyping are well-suited to programs of FCS's complexity. Such processes enable continuous feedback through intensive customer links in an iterative requirements-setting process, dynamically integrate knowledge, and lower the cost of future changes. The FCS LSI may benefit from exploring the use of flexible development processes in requirements setting, product design, and production.

As a private company, the LSI uses both financial and nonfinancial strategies to recruit, train, and retain a very experienced and talented team for the FCS project. Although there is a proactive mentor / protégé program, and significant value placed on retaining certain people for succession planning, there are challenges as talented individuals are promoted out of the program.[19]

CONCLUSION

Given the managerial capabilities of the LSI, our study developed the following conclusions. The LSI role is crucial to managing the complexity, uncertainty, and distributed nature of large-scale DOD systems integration programs and ecosystems such as FCS. For the customer, the LSI role will ultimately result in a higher "capability per unit cost" compared with traditional program procurement. This role becomes increasingly valuable as the systems "go live"—and as cost pressures and interoperability requirements escalate. Removing the LSI keystone role from the FCS ecosystem will result in a dramatic and chaotic reorganization.

FCS is a program of unprecedented complexity and uncertainty; as such, it is one of the most challenging product development efforts in history. The combination of rapidly evolving threats, constant injection of new technology, and high budget and political uncertainly will likely continue for the duration of the program. Based on our study of 31 megaprojects, the FCS LSI is implementing many best practices in the areas of premier program and ecosystem management. At the same time, the FCS LSI is being criticized in some corners for not locking in specifications and finalizing technologies before advancing the program. Such rebuke may misconstrue the very nature of the program's objectives and the types of challenges faced by a program of this scale.

APPENDIX: SUMMARY OF STUDY RESULTS

Our analysis of the 23 megaprojects found a highly significant statistical correlation between total management capability score and total program outcome score (figure 8.11). In general, the greater the overall management capability exhibited in aggregate across all three dimensions (program management, ecosystem management, process integrity), the more positive the overall outcome in aggregate across the four outcome dimensions (budget, schedule, performance, customer satisfaction).

It is interesting to look at the magnitude of the difference between top-performing programs and the worst-performing programs. The top quartile of performing programs scored an average 7.0 points (out of a maximum of 9) on the capability score, while the bottom quartile of performing programs scored an average of only 2.3 points.

Table 8.2 displays also a highly significant correlation between total outcome score and each of the three measures of program management capability. For the top quartile of performing programs, the average scores (on a 0–3 scale, with 3 representing the strongest expression of that capability) for program management, ecosystem management, and process integrity are 2.4, 2.0, and 2.6, respectively, while for the bottom quartile of performing programs they are 0.7, 0.8, and 0.8, respectively.

Figure 8.11. Total Management Capability and Total Outcome Score

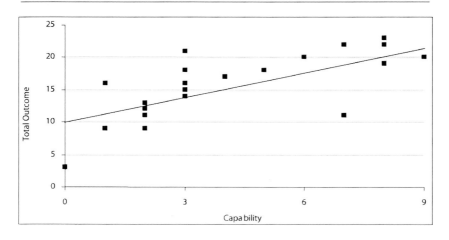

Table 8.2. Total Outcome to Three Management Capability Dimensions

	Unstandardized Coefficients	T-Score	Significance
Program management	2.413	3.511	0.002
Ecosystem management	3.505	3.918	0.001
Process integrity	3.155	4.207	0.000
Total capacity	1.381	5.365	0.000

NOTES

1. Alec Klein, "The Army's $200 Billion Makeover," December 7, 2007, *Washington Post*, http://www.washingtonpost.com/wp-dyn/content/article/2007/12/06/AR2007120602836_pf.html, p. A01 (accessed May 1, 2008).

2. The Army's modernization program, known as the "Future Force," is designed to make the Army lighter, more modular, and more deployable. It consists of 18 systems that include manned and unmanned ground vehicles, unmanned aerial vehicles, and sensors. These systems, which are connected by a network and, crucially, include the soldier, are frequently referred to as "18+1+1," or "18+network+soldier." Program Manager, Future Combat Systems, Brigade Combat Team, April 11, 2006.

3. Outcomes were measured against adherence to budget, adherence to schedule, fulfillment of performance objectives, and customer satisfaction.

4. The figures in the row labeled "average for large-scale projects" are based on quantitative studies of 47 civilian megaprojects from the following sources: Edward W. Merrow, *Understanding the Outcomes of Megaprojects: A Quantitative Analysis of Very Large Civilian Projects* (Santa Monica, Calif.: Rand Corp., 1988); Bent Flyvbjerg, Nitz Bruzelius, and Werner Rothengatter, *Megaprojects and Risk: An Anatomy of Ambition* (Cambridge: Cambridge University Press, 2003); and Alan McCormack and Kerry Herman, *The Rise and Fall of Iridium* (Boston: Harvard Business School Press, 2001).

5. Congressional Research Service, *The Army's Future Combat System (FCS): Background and Issues for Congress*, RL32888 (Washington, D.C.: Library of Congress, April 28, 2005), p. 9; Tom Phillips, *FCS Program Overview*, PowerPoint presentation, March 11, 2005, http://www.boeing.com/defense-space/ic/fcs/bia/050311_ovrview_forum.pdf.

6. John Viega and Gary McGraw, *Building Secure Software* (Boston: Addison-Wesley Professional, 2006).

7. The FCS development and manufacturing work is occurring in 33 states and 127 congressional districts. Tom Phillips, *FCS Program Overview*.

8. See Kim B. Clark and Takahiro Fujimoto, *Product Development Performance: Strategy, Organization, and Management in the World Auto Industry* (Boston: Harvard Business School Press, 1991), and Kim B. Clark and Steven G. Wheelright, *Revolutionizing Product Development* (New York: Free Press, 1992).

9. The "premier" mode of program management structure is also referred to in the literature as "heavyweight."

10. See Marco Iansiti and Roy Levien, *The Keystone Advantage: What the New Dynamics of Business Ecosystems Mean for Strategy, Innovation, and Sustainability* (Boston: Harvard Business School Press, 2004).

11. See Marco Iansiti, *Technology Integration: Making Critical Choices in a Dynamic World* (Boston: Harvard Business School Press, 1997).

12. See Clark and Fujimoto, *Product Development Performance*, chapter 9, and Clark and Wheelright, *Revolutionizing Product Development*, chapter 8, for background and discussion on the applicability of different program management structures to different program types.

13. This gave the program management organization visibility into the issues facing each group and provided the capability to uncover issues with critical implications for the overall project in the areas of systems integration, budget, and schedule. Program management organizations often face a fine balancing act between pushing problem solving down to functional groups and taking responsibility for issues with systems-level ramifications. Maintaining domain expertise in the program management organization establishes a forum for the program manager to determine the implications of any issues

across systems and then draw on the relevant expertise available in the group to identify and assign resources to resolving the problem.

14. More specifically, there is usually at least a government program management office and a prime contractor program management office. In some cases the situation is more complex, with multiple contractor program management offices involved and a joint program office (JPO) above them.

15. As strong as the capabilities are in this area, the LSI may have the opportunity to strengthen its role in simulation and test by deepening processes and capabilities in this area (e.g., drawing on the simulation and test practices for satellites, which have only one "shot" and therefore must test rigorously), particularly with respect to the complexity of such a distributed organization. See the "Process Integrity" section in this chapter for more discussion on the importance of simulation, test, and experimentation.

16. One program, for example, fielded a team of end users and put them directly in contact with designers and manufacturers in the supply chain, then kept that customer team involved through program deployment, ensuring that end-user requirements were accurately reflected in the design and operation of the product.

17. There may be additional opportunities for Boeing to establish processes and incentives across ecosystem members to "bubble" innovative concepts to the surface that would have tremendous value as part of FCS.

18. Going forward, there may be opportunities for the LSI to achieve greater efficiency through the use of incentives, to encourage not only compliance but active innovation in the area of similar standards. These efforts could be augmented by encouraging a culture of innovation focused on the overall FCS system, both at the corporate and individual levels, as discussed above.

19. There is an opportunity to establish methods to retain key individuals for continuity and retention of sticky knowledge through the program. Boeing also has the opportunity to deepen the use of iterative prototyping, simulation, and experimentation processes and capabilities in a virtual environment across multiple companies. Best practices from commercial software development projects might also be leveraged to support Boeing and its subcontractors in the development of the FCS software environment.

MANAGING GOVERNMENT EFFECTIVELY IN A COMPLEX ENVIRONMENT
INFLUENCING NETWORKS THROUGH THE NETWORK CAMPAIGN

W. SCOTT GOULD AND JULIE M. ANDERSON

As the federal government increasingly turns to the private sector for advanced products and services, agencies and companies are finding it harder to deliver high-quality results on time and within budget. Several factors are contributing to this trend. Between 1999 and 2007 the number of employees in the private sector working under contract to the federal government jumped 27 percent, growing from 5.9 million to 7.5 million. At the same time, the number of stakeholders involved in decisions around these contracts, as well as those who need to be consulted during delivery, has increased. Furthermore, large-scale government projects tend to encompass a wider range of concerns, cut across agency boundaries, and require a broader range of skills sets. Yet, fewer skilled government managers are available to plan, monitor, and evaluate this work as experienced managers continue to reach retirement age. Finally, the issues that government should address, such as terrorism or poverty, are becoming more complex, difficult to define, and hard to understand. Their underlying causes are more challenging to identify. The result is an increasingly complex environment in which government and business work to deliver outcomes through progressively more complicated projects.

In such a complex environment, government needs a new strategy to produce better results. Within this new strategy, government should be viewed more as a network than as a bureaucracy, with features from both. Networks exhibit "one-to-many" and "many-to-many" relationships between stakeholders in the system. These relationships are multifaceted and multidirectional, and it is often unclear who within the network has complete control while moving network objectives forward. It is easy to be confused by the distributed and overlapping roles and responsibilities among stakeholders within a network. At the same time the traditional decision rights and silo effects of bureaucracies further complicate the role of managers seeking clear direction. What emerges is a patchwork of conflicting incentives, misaligned systems, and overlapping objectives that thwart communication and collaboration between stakeholders in the network and undermine productive relationships between government and business.

This chapter describes a new strategy developed by the private sector to address the increasingly complex government environment. We review the principal factors that have led to this increased complexity and present a new strategy involving a Network Campaign (hereafter "the campaign") that will help government and commercial participants in the public sector navigate this environment and improve results through collaboration. Finally, we consider an example of successful implementation of this approach in order to illustrate its benefits.

THE PERFECT STORM: EMERGENCE OF THE NETWORK, PRESSURE TO PERFORM, AND RISING COMPLEXITY

Any successful strategy for dealing with complexity in the federal market should account for the network nature of government. The number and variety of stakeholders in the federal government network is large. This collection of stakeholders, joined by a common desire to influence decisions in the federal government, is the federal government network. Stakeholders include the government itself, which consists of political leaders and career employees from the executive and legislative branches at the federal, state, and local levels. Private contractors are also stakeholders, as are the nongovernmental organizations that deliver services on behalf of the government, including, for instance, the Red Cross. In addition, within the federal government network, there are external stakeholders that influence the government, such as the media, think tanks, academics, former government employees and leaders, constituents, and private sector lobbying initiatives. These government network stakeholders share information and shape, jointly develop, and sometimes implement public policy. Through these processes, they work to improve outcomes or to obstruct change. To work effectively in this context, one must know the key stakeholders in the government network and have an effective method for communicating and coordinating with them.

THE RISE OF THE FEDERAL GOVERNMENT NETWORK

This government network has assumed more roles and grown more powerful. Constituents are demanding a more efficient government capable of providing timely and effective policy outcomes. They have learned how to engage government more successfully using their representatives, lobbying groups, and the media. Information pertaining to government programs, though still difficult to acquire, is growing more available to a wider audience. This availability has helped network participants obtain performance data and use it to influence government, while holding it more accountable for its decisions and policies. As the economy transitions from labor-based production to knowledge-based, traditional hierarchical arrangements in both sectors have been marginalized, thus empowering a variety of employees within government as well as citizens and interest groups to form networks, communicate, and speak up to influence government.

The network also helps stakeholders create informal organizations more rapidly and assists them in achieving their goals by using Web technology to leverage easily accessible, low-cost information to disseminate their positions more widely and rapidly. Stakeholders can mobilize and act more quickly today, and they are clear on what they want. For example, in September 2006, conservative bloggers banded together to expose a senator who had placed a secret "hold" on a piece of legislation that created an online, searchable database of all federal government grants. The legislation was intended to make the budget more transparent and reduce pork barrel spending. Bloggers called each senator's office to narrow down the list of who was holding back the legislation—a long-standing tradition within the Senate. They identified Senator Ted Stevens (R-AK), the chairman of the Appropriations Committee, and made his identity public. Senator Stevens was forced to release the hold, and President Bush signed the bill into law in early October 2007. With the ability to access information quickly and coordinate their strategy, the bloggers were able to influence policy change within a relatively short period of time.

Overseeing the collective enterprise of outsourcing is a network of government officials, corporate board members, shareholders, public sector unions, Congress, courts, and a variety of nongovernment stakeholders, including the media. And, government has developed a set of inter-agency and intra-agency management strategies to solicit input from, consult with, and respond to a large number of participants. The result is a system with an extraordinary number of players, many capable of holding up progress. In government, many people need to say "yes" for an agenda to succeed, but only a few need to say "no" in order for it to fail.

INCREASED PRESSURE TO PERFORM

At the same time that the government network has become more prevalent and more powerful, it is under increased pressure to perform exceedingly well and at lower cost. This can be witnessed in the ongoing shift of burden from federal to state and local programs in the last decade as federal funding for some programs, such as Medicaid, has been reduced while demand for these programs generally has increased. In addition, other factors have added to the pressure to perform, such as increased government spending, decreased receipts, the consequences of mismanaging expectations or excluding important stakeholders in the network, and a rise in the total number of employees in the blended government workforce. Combined with the cost of the war on terror, this trend has placed new and very significant pressures on all government agencies to do more with less. There is general agreement both inside and outside government that this pressure will grow more acute in the years ahead.

This pressure to perform remains even when government employs the routine process of solicitation, proposal, review, and contract when faced with a specific challenge. It is yet more difficult when the government either does not understand a challenge well or is unfamiliar with new and innovative approaches to solving the problem used by the private sector. In this case, the federal government can miss an opportunity to address a critical challenge or may choose a suboptimal solution.

Finally, consumers of public sector services witness higher-quality customer service in the commercial space. Knowing the government can do better, they are less willing to forgive poor service, whether they are renewing a driver's license or receiving emergency aid in a natural disaster.

THE RISING COMPLEXITY OF GOVERNMENT PROBLEMS AND ENVIRONMENT

The country is facing increasingly complex problems. These problems, which are multifaceted and multilayered, involve a web of personal and organizational relationships and touch on a greater range of desired outcomes. Further, these problems require a greater level of coordination to address—coordination between departments in the federal government, among state and local agencies, with foreign governments, and across sectors.

Moreover, solutions must be delivered by organizations whose missions are more difficult and riskier to fulfill than in the past. Highly interconnected systems like global trade, international finance, and worldwide communications have become more vulnerable. The more tightly coupled these systems become, the more susceptible they are to failure because a problem in any single node is linked to the entire system. In the government setting, a risk in one part of the network, such as a failed initiative in an agency, could undermine the prospects for another agency to accomplish its mission in another part of the network.

To understand this complexity, participants should think of the players in government as part of a network as much as a traditional bureaucratic organization marked by rigid structure and procedures. Government managers should be mindful that agencies overlap in their authorities within the federal government and between state and local, have unclear authorities open to appeal at all times, and are defined by shared interest and risk. Government managers can employ different tactics and make decisions in another way, knowing that the intended results of their project will be subject to indirect, distributed, rapid, and wide-ranging influences in the government system.

Policy formulation in a public setting is more challenging because of the number and variety of stakeholder interests. Problem solving is more intricate and nuanced in the public policy network. Public sector leaders can no longer cooperate and communicate with just a few key nodes in the network where power is concentrated. They should engage with a wider range of actors, each with the power to block, shape, or delay the desired result.

For these reasons, public sector organizations are under greater pressure to solve more complex problems in less time, using fewer resources with lower margin for error. And, they must solve them in a way that satisfies the preponderance of stakeholder needs in the federal government network.

COLLABORATING WITH THE PRIVATE SECTOR TO MANAGE COMPLEXITY

To meet these challenges, government has turned increasingly to the private sector to obtain specialized products and skills and to achieve related cost efficiencies. During the past 10 years, the government services industry in the United States has expanded at an average compound annual rate of 6 percent. The result is an increasingly multisector workforce—that is, one composed of a proportionally smaller core of full-time public sector employees working in an ever larger network of private sector service providers. Faced with different incentives, these two parties have struggled to align toward common objectives. The public sector has struggled to stay current, particularly with technology, because of a chronic lack of investment in training for civil servants. Meanwhile, the private sector has found it challenging to coordinate with a government that manages some programs poorly, requires complex procurement rules, and is unable to define requirements in a way that is suited to this procurement environment.

However, there are several techniques for dealing with complexity and delivering for results. First, managers are working to develop relationships that can cope without full information and an imperfect alignment of interests. Differences with respect to incentives, decision rights, and knowledge mean that the management challenges are broad in scope, layered in meaning, and nuanced in implementation. The contracts that support these arrangements should be flexible and more outcome-focused.

Second, managers are learning to make better trade-offs between control and coordination costs. The traditional method of instructing subordinates through control mechanisms is not effective in a networked, complex environment. Instead, managers can provide guidance to another organization or work unit about a particular problem. But this requires a great deal of knowledge, effort, and precision to communicate the specific tasks that should be performed by that group. The most effective ways to lower these coordination costs is to communicate objectives clearly, create a set of conditions that enable the work to be produced in a results-oriented process, and anticipate problems created inadvertently with other aspects of the complex solution.

Third, public sector managers are using change management techniques suited for government to incorporate new business processes and adjust to new performance goals. As these managers move their organizations from the current state to the new one, they are taking into account issues of culture, psychology, governance, business process, and technology. Managers are spending extraordinary time and effort convincing stakeholders of the benefits of the new approach by answering the key question, "What's in it for me?" They persuade stakeholders to do things in a new way by showing that it has been done before and by controlling risk through an incremental, spiral approach. In this way small, tightly defined projects with very specific, measurable goals allow managers to learn from previous processes and refine the level of quality before repeating the process. Similarly, component-based analysis allows managers to create an analytical frame through which they can view a complex process as a whole and use this vision to better understand the parts. The challenge with this type of analysis is knowing when it should be employed and when it could add to the project's overall risk if employed incorrectly.

Fourth, managers are using modeling and simulation as a tool to manage complexity. Computer models of business processes and their associated technology, human capital, and technology constraints allow users to assess quickly the likely outcome of a particular set of actions. This approach enables managers to bench test changes in policy and business practice in order to learn the likely outcomes before implementing the actual decisions.

TOWARD A NEW STRATEGY

Over the decades, government organizations have articulated their needs for consulting and systems integration services by issuing Requests for Proposals (RFPs) and operating a competitive bidding process for qualified contractors. Companies both large and small have learned to see their world of contract capture through a tightly focused lens. The public sector divisions of firms have adopted an argument that a narrow, laser-like approach to a particular bid is the best way to win and deliver value: respond to the RFP and address only the stated requirements. But this approach no longer reflects the current reality of most government deals, where the people with the money need to know that the supplier has a complete understanding of the whole environment and where the problems they are encountering are new, more challenging, and higher risk. This is not to argue that companies that take a traditional approach to meeting the stated needs of governments will not continue to do well or that governments cannot manage traditional projects effectively, but rather that large projects in which government and industry managers work together to understand needs, shape requirements, and deliver results will produce even more effective outcomes and greater value for constituents and taxpayers.

THE NEED FOR A NETWORK CAMPAIGN

Private industry has known for a long time that a coordinated sales process delivers better outcomes in the commercial sector because it incorporates the interests of multiple stakeholders in the business development and sales process. After all, a company that has better intelligence than the competition, fully understands the needs and expectations of everyone involved in the process, and can articulate just why everyone will win with their deal is going to be in a much better position to sell successfully and deliver. This same concept can be applied to managing large projects in a networked public sector environment with government and private sector participants. A process that allows government managers to understand the needs of all stakeholders and incorporate these priorities into their decisionmaking is more likely to result in agreement and yield better outcomes.

Yet, the public sector contracting world introduces two wild cards—governments and taxpayers. The right deal is not just the least expensive, it is also one that delivers real value that benefits everyone and can be sold as such by the different stakeholders. In other words, the supplier must have sufficiently deep understanding that a successful contract bid can be sold both up and down the value chain as the right answer for all involved. For example, a consulting and technology firm hired by the federal government to develop and implement electronic health records for every member of the military is more likely to create an effective system and processes if it fully understands how the individual, the health care provider, and the payer will use the end product.

In theory, the contractor should be one of the partners in executing public policy by delivering a product or service that mirrors the public policy requirements. For example, the Department of Defense (DOD) changed its relationships with its suppliers in its Joint Direct Attack Munition (JDAM) program beginning in the late 1990s. Program managers created a long-term relationship with suppliers and provided input about the requirements and execution of the program. By applying prioritized criteria, the DOD reduced the number of suppliers, streamlined buying processes, saved administrative costs, and maintained competition among bidders.[1] All too often, however, the contractor only sees the contract in terms of dollars and cents, another contribution to the bottom line. Companies and government organizations that adopt transactional approaches can be successful, but they leave a large amount of value to taxpayers, constituents, and shareholders on the table. This short-term approach to managing contracts can limit the value delivered to government agencies and the constituency groups they serve through unclear requirements and a gap between public sector needs and private sector capabilities.

As money gets tighter in the years ahead, the government will turn increasingly to the private sector for outsourced services. Companies that fully understand their clients and their interests, priorities, and incentives will produce more effective outcomes, benefiting the public good and their own bottom line. Government organizations that fully understand their private sector partners and incentives will manage contracts more effectively, resulting in greater value to their stakeholders, taxpayers, and constituents. Although outsourcing will increase, so will congressional oversight. Ultimately, agencies and companies that work together not only to budget and deliver on-time results but to keep a careful eye on the needs of clients and congressional stakeholders will be successful.

This kind of Network Campaign will always be an iterative process like all forms of organizational development. We expect that one contract will inform the next—one campaign will deliver lessons that will be incorporated into the next phase. By using the campaign to its full capability, the result will be a honed and highly efficient process that will make the corporation agile, informed, and completely responsive.

WHAT MAKES FOR A GOOD CAMPAIGN?

An efficient and effective campaign in the federal market is designed to create a set of permissions among decisionmakers in the network in order for an action to take place. Acting in concert, the branches of government make policy decisions and ultimately shape procurements through a distributed decisionmaking process. It takes many people to say "yes" in government and only a few to say "no." To get an idea across in government, many people have to hear it and accept it. Therefore, the campaign should include the following components:

- **Identification of key influencers.** Key influencers, who may range from a congressional aide to the actual decisionmaker within an organization, are individuals with the power, the means, and the desire to influence an issue. For companies conducting business in the global public sector, these stakeholders include members of various bodies, such as Congress; the executive branch, including the White House; federal government agencies; various regulatory groups; and their global counterparts in parliaments, ministries, and offices of heads of state. In addition, stakeholders outside of government that wield influence include media; opinion makers; nongovernmental organizations such as think tanks, universities, nonprofit, and advocacy organizations; and government relations firms. Campaigns focused on the state and local levels of the public sector should target comparable organizations. Additional research can reveal specific people, personalities, and connections that further hone targeting strategies for key influencers.

- **Coordinated action plan.** Today, all organizations that deliver services and products are highly networked with many-to-many relationships. The delivery needs to be carefully coordinated through a process of continual innovation designed to create good decisions for the company and its clients. This action could be passage of legislation, issuance of regulations that govern eligible methods to spend federal dollars, or a decision to use private contractors to design and develop new ways of delivering services to constituents. A coordinated set of actions within the network raises the likelihood of success for business development activities conducted by the firm.

- **Effective marketing.** The organization's marketing effort can set the context for discussion of a specific policy issue when influencing policymaking activities within legal and ethical bounds.

- **Contingency management.** Because situational awareness and intelligence are never good enough, the campaign must include processes that allow for managers to quickly incorporate new circumstances.

WHAT MAKES THE CAMPAIGN UNIQUE?

The campaign should not be confused with a more traditional marketing campaign that generally seeks to position activities such as advertisements and slogans to describe a new product or service. Instead, this approach employs a multifaceted strategy and an array of tactics to communicate and reinforce key messages, influence decisionmakers, and facilitate the adoption of new ideas. This approach is more similar to efforts known in political and government circles: an organized effort to influence the decisionmaking process within a group or to incite action. The most effective approach is multivariate—an approach suited to the layered, overlapping, multi-stakeholder government environment. It can require influencing multiple nodes in the network simultaneously. And it can begin influencing activities at a number of nodes in the network at the same time, subsequently driving toward the same objective.

The multivariate nature of the campaigns is intended to capitalize on policy windows and create new policy windows: crucial points where public opinion and legislative and executive branch attention suddenly move in a new way. External events and circumstances sometimes create unique conditions in which decisionmakers are motivated to act on a certain issue and determine, in part, business opportunities for companies serving the public sector.

For example, the controversy surrounding the spring 2006 bid by Dubai Ports World to purchase port operations at six U.S. ports created an election-year opportunity for congressional members to push through port security legislation—an issue to which constituents and the media had previously paid little attention. Companies targeting public sector clients in the homeland security market had a vested interest in congressional and executive branch action on this issue. They had an opportunity to shape legislative language and regulations that may induce demand and/or create more market certainty to support the selling of their products and services.

However, the conditions creating policy windows are often unpredictable in a distributed environment. Organizations should plan for contingencies and execute carefully to create opportunities to take advantage of these windows.

CREATING THE CAPABILITY: A FIVE-STEP PROCESS

All campaigns have five basic steps: generating insight, creating leading ideas that reflect stakeholder needs, building influence, making a

Figure 9.1. The Five Steps of a Campaign

connection to organizational objectives, and coordinating across many units and teams in the organization. Our experience suggests that each one is important in creating the capabilities needed to plan and execute a campaign effectively and needs to be combined with the components of the campaign outlined above.

1. **Understand the Environment.** An organization should conduct research on the key issues and areas of interest of public sector decisionmakers and stakeholders in order to select and prioritize the topics on which it will focus. This step can be thought of as an assessment of client needs and stakeholder (market) priorities.

2. **Envision the Future.** An organization should marshal its resources to develop new and innovative ideas about how to address the prioritized issues. Solving public sector problems in new ways requires innovation. Leading with ideas and then building the infrastructure to support them will enable the organization to move ahead of competitors.

3. **Generate Influence.** The ideas should be disseminated through multiple channels to create the set of permissions for decisionmakers to act. Adoption of the ideas by multiple stakeholders will begin to shape leaders' choices.

4. **Coordinate Opportunity Creation.** The feedback from stakeholder interactions and new information from decisionmakers should be incorporated into the process in order to develop the new services and products required to enact the leading ideas.

5. **Support Execution.** The leading ideas should be connected continually to implementation of the services and products at a project site. The lessons learned from implementing services to meet client needs should be integrated into organizational strategies and tactics moving forward.

Effective campaign management links these steps together in a coordinated and iterative process. It ties the process to the rest of large private and public organizational systems—for example, sales, marketing, governmental relations, media outreach, policy development, program oversight and evaluation, and budget. It maximizes consistency while minimizing risk. As any experienced executive knows, it is impossible to optimize fully both consistency and risk. Therefore, the campaign approach will typically involve a trade-off where one component is maximized while the other will operate within an acceptable range.

This campaign also provides a central point for feedback, which enables the organization to adjust its campaign strategy and tactics based on new information obtained through campaign activities. Finally, campaigns employ several unifying principles, including integrated deployment of capabilities and agile market response.

MAKING THE CAMPAIGN WORK

Managing a campaign in large, distributed government agencies and contractor organizations is challenging. Pursuing a campaign requires the organizations to tighten their internal coordination in order to change the innovative process effectively. This challenge is tackled by establishing key roles, creating regular communications mechanisms to support coordination, securing adequate funding, and creating a steering committee to provide input and feedback about key decisions. The organization should rely on an *executive sponsor* to support integrated procedures, a *thought leader* to develop new insight about the market and innovation, and a *campaign manager* to formulate strategy and tactics and drive execution of the planned activities. In addition, the campaign manager will need to coordinate across a wide variety of players from the media, marketing, sales, and research units of the organization.

The campaign manager should also establish a small advisory board composed of executives (key stakeholders) who can provide guidance and feedback about key directional decisions. Each advisory board member should have a vested interest in the outcomes of the campaign. For example, advisory board members could be asked to contribute to the funding of the campaign from their own operational budgets. The advisory board should meet frequently—for example, once a week—to harmonize priorities, offer help in clearing organizational barriers, and provide feedback on strategies and tactics before approaching clients. In a government project environment, the advisory board can be composed of government and private sector stakeholders so that the priorities of multiple organizations become aligned to maximize coordination and effectiveness of outcomes. The campaign manager held weekly conference calls with stakeholders to maximize the number of people receiving the same information in a large organization where it is easy to tune out e-mails. Finally, a formal system for measuring performance should be established.

The U.S. Congress typically takes many years to consider legislation. In the federal government, environment issues evolve slowly with multiple stakeholders influencing, shaping, blocking, or changing the proposed laws. The thinking of decisionmakers also evolves as their internal priorities and outside pressures change over time. Campaigns developed around existing and anticipated client needs require organizations to focus on the long term. Campaign strategies should be long term and flexible to reflect this evolutionary environment. Expected payoffs from campaign strategies and tactics may take as long as 24 to 36 months to be realized.

THE PUSH-PULL APPROACH

Campaigns use both "pull" and "push" methods to achieve their objectives. For example, environmental advocacy organizations have focused on persuading the federal government to address the causes of global warming during the last 15 years. These organizations have relied on building relationships with congressional decisionmakers in order to share the results of scientific research, ask them to enact policies that support efforts to reduce global warming, and encourage the American people to become involved in the cause. At the same time, leaders within executive branch agencies such as the Environmental Protection Agency are implementing their own policies to carry out the mission of their organization. Combined with the external factors of higher fuel prices, these deliberate strategies and tactics have helped shift public opinion about the importance of reducing greenhouse gas emissions and drummed up support for a more active government role in achieving this objective. As a result, their efforts have created new markets for alternative energy sources.

A pull method is one where the government agency's interest, power, and independence dominate. The pull method drives a solution

> **Case Study:** IBM Employs Network Campaign in Global Movement Management Market
>
> In 2006, IBM embarked on a plan to accelerate growth in revenue for its U.S. government services business using the campaign process. (At the time, the U.S. government services market represented $68.5 billion; the global public sector services market is much larger at $134.7 billion.[1] The disciplined and repeatable process of the campaign led IBM to develop innovative and visionary ideas and communicate them effectively to clients. IBM selected the global supply chain market in which to apply this approach.
>
> To cultivate these ideas in the public sector market, IBM developed Global Movement Management (GMM), a comprehensive framework for securing the global trade and travel system against disruptive threats. The framework was described in a series of white papers that avoided mention of any proprietary technologies or particular industry solutions. The white papers served as the foundation for shaping client and stakeholder choices.
>
> IBM subsequently engaged in an active 24-month campaign to persuade policy and opinion makers, constituencies, and potential clients of the benefits of GMM. The campaign employed a series of one-on-one conversations, published articles and opinion pieces in leading publications, conferences and media events to convey its message. IBM's efforts not only provided its government clients with a new understanding of how to secure the United States and its economic interests through improved policies and programs, but they also produced new revenue opportunities in the form of competitive government contracts. The campaign helped IBM identify a need, develop the thought leadership, communicate with the different constituencies, create a new market, and then compete in that market. In addition, the nature of the conversations with clients around the campaign material created the opportunity to listen for new insights, discuss problems from a fresh perspective, and qualify new business prospects. The campaign led to $450 million in new business and contributed to a recent $3.6 billion contract win.
>
> ---
>
> 1. Gartner Dataquest, Market Statistics, "Industry IT Services Worldwide Market Share: Database," April 2007.

to a public sector leader who wants to forge ahead with an idea that would ultimately induce demand. A push method, on the other hand, employs public opinion and congressional interest, power, and control to encourage recognition of a problem and possible solutions along with permission to act. In this instance, a campaign builds pressure for change from outside an organization. This can occur during a time of national emergency, a critical transition phase, or business as usual. A combination of methods involving both pull and push approaches overcomes inertia on the side of both the agency and the external stakeholders and encourages them to deal with a problem. This is the most common and most effective approach in a multi-stakeholder environment. Organizations conducting a network campaign have an interest in using both methods to influence possible business outcomes in a multivariate environment.

The campaign employs a series of shorter-term tactics consistent with strategies and aimed toward accomplishing the objectives of the campaign. Thought leaders working with the campaign through lobbying, earning press attention, advertising, marketing through events and written materials, and making speeches support the implementation of campaign strategies. To execute these tactics, managers must coordinate across the organization to maximize opportunities, harmonize priorities, amplify messages, and represent a balanced perspective on behalf of the organization.

Campaigns can be managed as a portfolio with mechanisms for adding, modifying, and culling underperforming campaign efforts. A division of a large organization could develop and manage multiple campaigns to penetrate different markets simultaneously. A portfolio effort requires identifying emerging trends, adding new campaigns based on these trends, evaluating campaigns and their effectiveness in the context of these trends, removing underperforming campaigns, and allocating budget resources to priority campaigns.

CONCLUSION

In our experience, the Network Campaign is a vital component for successfully managing complexity in the public sector environment. Government agencies can use it to gain greater understanding of their requirements, manage multiple stakeholders of large, complex information technology (IT) systems, and shape the activities of those who will operate and maintain the system for mission success. Similarly, the private sector can use it to identify potential revenue opportunities in the public sector and then convert that potential to the bottom line. It is no longer good enough for sales organizations to work in isolation with short-term goals that misread the long-term strategic implications for their own enterprise and for their client in the public sector. Nor is it effective for government agencies and their large-scale project managers to apply top-down management approaches in a networked environment.

The last few decades have seen a significant shift in the way the public sector does business—a change driven by technology, the complexity of new networks, greater oversight, and the need to make every dollar spent deliver ever-higher value. All these changes will continue to affect the market in the years ahead, especially if, as expected, local, state, and federal budgets continue to be squeezed. New approaches to manage complex, large-scale IT systems implementation will continue to be developed. We have argued that this is particularly true with respect to the process by which both government and industry partners identify and articulate proactively the needs of the public sector client, anticipate future demand, and then position to meet those needs.

An effectively designed and implemented Network Campaign allows government organizations to develop requirements and manage program implementation for success. It sets the conditions for companies to supply goods and services to the government with better overall results. This concept, though simple, requires for successful execution a structured and flexible process to meet the unique and complex demands of networked government. This is exactly what the campaign delivers.

NOTE

1. U.S. General Accounting Office, *Best Practices: DOD Can Help Suppliers Contribute More to Weapons System Programs,* Report to the Subcommittee on Acquisition and Technology, Committee on Armed Services, U.S. Senate, Report number GAO/NSIAD-98-87 (March 17, 1998).

HUMAN CAPITAL FOR COMPLEXITY

DAVID H. DOMBKINS

Over many millennia, humans have successfully delivered astonishing projects and programs. Humans have always pushed their competency limits and pursued visions in advance of technological and social capabilities. So what is different now? The answer is that many of today's projects and programs are more than complicated—they are truly complex.

The differences between complicated and complex programs are not readily understood. Complicated programs are relatively common and are usually delivered by decomposing the program into subprojects and then resolving interdependencies (integration) between subproject boundaries using systems engineering. To many, complicated programs will seem complex. Complicated programs, though usually very large, can be defined in scope to a high degree of accuracy at inception and throughout the design phase. This is in stark contrast to complex programs, which are described as follows:

- open and emergent systems characterized by recursiveness and nonlinear feedback loops. Their sensitivity to small differences in initial conditions and their emergent nature significantly inhibit detailed long-term planning for these programs;
- systems that exhibit dynamic complexity, where parts of the system can react and interact with each other in different ways (a chess game);

- systems often implemented in highly pluralist environments where multiple and divergent views exist; and

- "complex evolving systems" dominated by double-loop learning. More than "complex adaptive systems" that simply adapt to their environment, they change the rules of their development as they, over time, evolve with and are open to that environment.[1]

A significant shift in mindset is required to move from a paradigm of certainty to a paradigm of complexity. This change in paradigm is fundamental to understanding and implementing complex programs. In many boardrooms and general management business schools, traditional project management and complicated program management are viewed as operational-level activities. This is in stark contrast to complex program and portfolio directors who are senior executives, possess unique skills, and work in chaotic environments to deliver nonstandardized and emergent strategic outcomes.

This chapter proposes the following:

- Projects, programs, and portfolios are best understood when viewed as life cycle–based systems.

- General management and systems thinking are essential aspects of both traditional project management and complex program management.

- Project management is a continuum that

  — at one node has traditional project management, or TPM, with its philosophy, strategy, organizational architecture, methodology, tool set, process, and contract all firmly based upon certainty;[2]

  — at the other node has complex program management, or CPM, with its philosophy, strategies, organizational architecture, methodology, tool set, processes, and contracts all firmly based upon complexity;

  — at the intersection between traditional project management and complex program management has the cusp area (called complicated program management) where the project manager is highly competent in traditional project management, possesses an awareness of the complex program management paradigm, and has developed competencies in a range of the complex program management competencies; and

— has three distinct types of project managers:

— traditional project managers—vocational training and accreditation;

— complicated program directors—tertiary training and peer-based accreditation; and

— complex program and portfolio directors—tertiary education, mentoring, and peer-based accreditation.

- A formal career pathway for project managers needs to be established that combines the International Project Management Association (IPMA) traditional and complicated project manager certification and the Complex Program Institute (CPI)[3] complex program and portfolio director certification; and

- There is an urgent need to fast-track the development of a cohort of complex program directors to deliver complex environmental, sustainability, social, health, change, ICT, infrastructure and defense projects.

PROJECT TYPOLOGIES AND COMPLEXITY: A DIFFERENT PARADIGM

The key to understanding the differences between the three types of project management (complex, complicated, and traditional) is to examine their different philosophical bases (see table 10.1).

The differing philosophical bases lay the foundation for the competencies, methodologies, tool sets, processes, and contractual models for each type of project management.

To enable organizations to differentiate between traditional, complicated, and complex, a reliable and valid tool set is required to categorize projects/programs. The initial methodology for classifying projects by type was the WHOW matrix (see figure 10.1).[4] It provided a subjective means of classifying projects/programs according to their certainty in both scope (what) and delivery methodology (how).

The Project Categorization Framework (PCAT) builds on the WHOW matrix to provide a valid and reliable methodology to cat-

Table 10.1. Philosophical Bases of Project Management Types

	Traditional	Complicated	Complex
Philosophical base (paradigm)	Certainty	Certainty and pluralism	Uncertainty, pluralism, chaos, and emergence

Figure 10.1. The WHOW Matrix

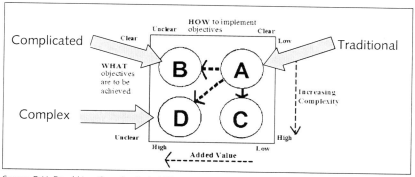

Source: D.H. Dombkins, *Complex Project Management* (Charleston, S.C.: BookSurge, 2007), p. 294.

egorize projects/programs (see table 10.2). In addition to categorizing projects/programs by type, PCAT defines the appropriate level of competency certification—that is, the appropriate tools, processes, strategy, and contract. PCAT uses the International Project Management Association (IPMA) project manager certification levels for traditional projects and complicated programs, and the Complex Program Institute, or CPI, certification levels for complex programs.[5]

The assessment criteria and weightings used in PCAT categorization are as follows:

LEVEL OF EMERGENCE

The project/program is a journey driven by a vision with high uncertainty in scope definition. Systems function as a whole, so they have properties above and beyond the properties of the parts that comprise them. Known as emergent properties, they emerge from the system in operation. One cannot predict the behavior of an emergent system from studying its individual parts. The level of emergence is a measure of the following:

Table 10.2. The Project Categorization Framework

PCAT Number	Project Description	Project Management Competency	What	How	WHOW Type	PM Certification
PCAT 1	Highly complex project	Complex	Unclear	Unclear	D	CPI complex portfolio director
PCAT 2	Complex project	Complex	Unclear	Unclear	B/D	CPI complex program director
PCAT 3	Traditional project with highly political environment	Complicated	Clear	Unclear	B	CPI program director IPMA level A
PCAT 4	Traditional project	Traditional	Clear	Clear	A	IPMA level B
PCAT 5	Minor works	Traditional	Clear	Clear	A	IPMA level C

Source: D.H. Dombkins and P. Dombkins, *Contracts for Complex Programs* (Charleston, S.C.: BookSurge, 2008).

- the scale of strategic change,
- the depth of cultural change, and
- the level of technical emergence in the project.

Increasing either the level of emergence to be realized and/or the level of integration required increases the level of project complexity and therefore the level of project risk.

INTERNAL SYSTEM COMPLEXITY

- Team complexity—a measure of the complexity of the internal architecture of the project team and the maturity of the project team in this type of project/program.
  - Technical difficulty—a measure of the novelty of the project/program and inherent complexities that arise from technical undertakings such as conflicting user requirements, integration with supra systems, architecture, design and development, assembly, technical emergence, incremental/modular builds, integration, and test and acceptance.
  - Commercial—the level of usage of relational performance based, phased, and layered incentive-driven contracting arrangements and the complexity of the commercial arrange-

ments being managed, including the number and level of interdependent commercial arrangements.

EXTERNAL SYSTEM COMPLEXITY

- Stakeholder complexity—a measure of the complexity of the stakeholder relationships. It includes the number of stakeholders, the level of alignment versus pluralism, cultural diversity, and geographic dispersal.
- Schedule complexity—a measure of the inherent complexity arising from schedule pressures on the project. The project/program is delivered using wave planning, and is subject to competing and conflicting priorities.
- Life cycle—a measure of uncertainty arising from the maturity of the delivery organization and the environmental maturity within which the project/program will be operated, supported, and sustained.
- Costs—factors include requirements development (empirically 6 percent–10 percent of acquisition cost) and lifetime operating, maintenance, and support costs, as well as asset management and periodic upgrading (empirically 3–4 times acquisition cost).

PROJECTS/PROGRAMS AS SYSTEMS AND LIFE CYCLE PHASES

Systems thinking (ST) provides a continuum of views and metaphors to provide insights into different types of projects/programs: at one end are traditional positivist approaches and at the other are anti-positivist approaches. The anti-positivist approach to ST is referred to as complex ST (see figure 10.2).

A key aspect of this contingency approach is that, depending upon a project/program PCAT categorization, different systems thinking methodologies should be used.[6] The use of ST's multiple views yields greater insight and more practical understanding, though the multiple views may contradict one another and do not provide certainty.

Complex ST looks at projects and programs holistically with the project/program in context of its supra system rather than in isolation from its environment and with artificial boundaries. This is significantly different from traditional ST where artificial system boundaries are

Figure 10.2. Complex Systems Thinking

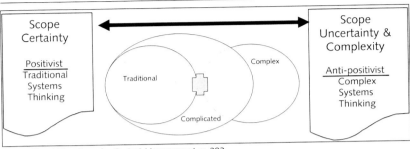

Source: Dombkins, *Complex Project Management*, p. 293.

established. Achieving this holistic view requires project management to use a new way of thinking. Traditional positivist project management methodologies and tools logically break down projects, organizations, and issues into their constituent parts, analyze those parts, and then reassemble them. This logical approach is limiting through its failure to address the interaction and synergy between constituent parts and between the project and its environment. The interaction and synergy between the elements within a system and the interaction of that system with its environment are the critical issues—not how the parts of that system operate in isolation.

Table 10.3 compares the three archetype project/program types and their respective system views.

In complex programs, the planning process is usually recursive and nonlinear rather than linear. Wave planning plots nodal points for gathering information, design, and implementation, allowing nonlinear and recursive patterns to be portrayed in a linear model. The program outcomes are usually delivered in modular (wave) builds (see figure 10.3). Figure 10.4 shows how wave planning can be applied to systems of systems (the seven core elements of the U.S. DOD system of systems guide).[7]

COMPLEX PROGRAM DIRECTOR COMPETENCIES

The three archetypes of project management are a career pathway build, with complicated building on traditional and complex building on complicated. A project management career pathway arises from integrating the dominant CPI and IPMA certification frameworks. However, given the philosophical and system differences across the three archetypes, it is appropriate that there are significant differences in the competencies required for each archetype.

Human Capital for Complexity **135**

Table 10.3. Three Project/Program Archetypes and Their System Views

Project Archetype	Traditional	Complicated	Complex
Nature of internal project system	Closed	Closed	Open (evolving and emergent)
Relationship with external environment	Limited interaction with the environment (artificial boundaries established)	Open and ongoing interaction with the environment managed by an environmental management system that protects the internal system	Ongoing interaction with the environment. The internal system and the external environment co-evolve.
Nature of external system	Closed	Open	Open (evolving and emergent)
Appropriate systems tools	System dynamics for internal system	System dynamics for internal system environment: • Strategic assumption surfacing and testing; and/or • Critical systems heuristics	Internal system and environment: • Variable systems dynamics; and/or • Interactive planning; and/or • Critical system heuristics
Appropriate competencies	Traditional project management for the internal system	Traditional project management for the internal system. Complicated competencies for the management of the external environment	Complex program management for both the internal system and the environmental management system
Appropriate contracts	Traditional scope and lump-sum fee	• Traditional scope and lump-sum fee with a partnering overlay; or • Alliancing	Governance contract using wave planning [a, b]

(continued next page)

Table 10.3. Three Project/Program Archetypes and Their System Views *(continued)*

Project Archetype	Traditional	Complicated	Complex
Life cycle	Traditional projects follow a linear phase model—concept, design, and systems engineering, implementation and verification, finalization and commissioning. The project life cycle of traditional projects is usually limited to project implementation.	Complicated projects and programs are often better viewed as programs made up of multiple interconnected traditional projects.	Complex programs are often non-linear and recursive in their nature. The phases of complex programs are discovery (including systems thinking, systems architecture, concept, design, and systems engineering), modular implementation and verification, double-loop learning,… repeated. The program life cycle for complex programs covers the whole of the program life from inception to disposal.
Project architecture[c]	Fully integrated with projects with predefined minimal external integration	Integrated across subprojects with pre-defined minimal cross-project/external integration	The functional, physical, and operational systems design that supports emergence and satisfies strategic whole-of-life and operational concepts and outcomes, from which flows the concept design and the development of component-level requirements

a. D.H. Dombkins, *Complex Project Management* (Charleston, S.C.: BookSurge, 2007).
b. D.H. Dombkins and P. Dombkins, *Contracts for Complex Programs* (Charleston, S.C.: BookSurge, 2008).
c. Project architecture is now recognized in the U.S. DOD's *Systems Engineering Guide for Systems of Systems* as a key determinant in the effectiveness of the system-of-systems to deal with emergence and its robustness. System-of-systems need to be proactively managed as a journey, using wave planning.

Human Capital for Complexity **137**

Figure 10.3. Modular/Incremental Build Using Wave Planning

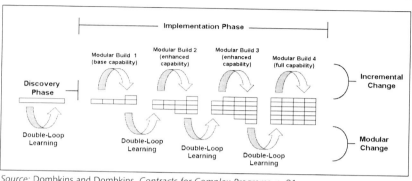

Source: Dombkins and Dombkins, *Contracts for Complex Programs*, p. 91.

Figure 10.4. Wave Planning as Applied to Systems of Systems

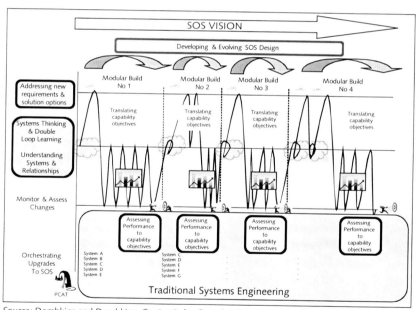

Source: Dombkins and Dombkins, *Contracts for Complex Programs*, p. 94.

Figure 10.5. Scope of Competencies for Project Managers

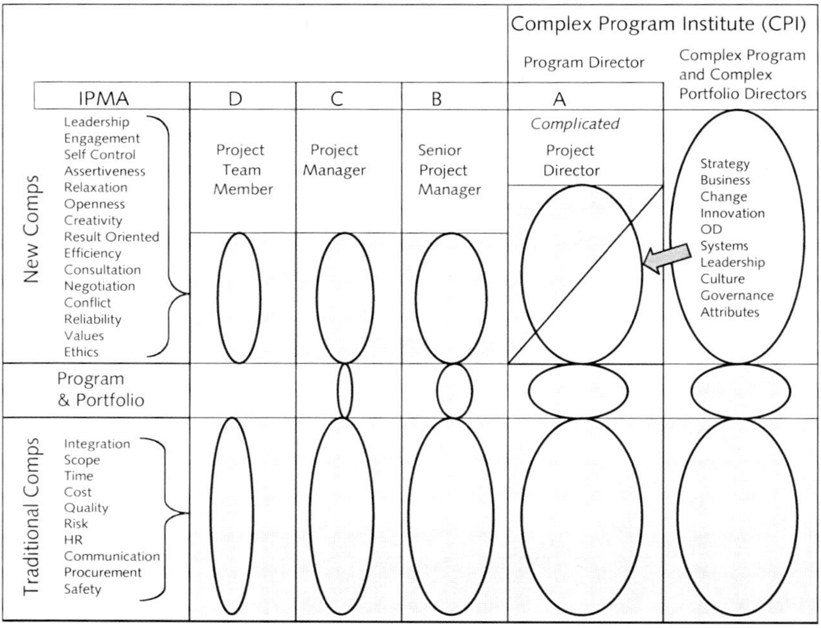

		IPMA	D	C	B	A	Complex Program Institute (CPI)
						Program Director	Complex Program and Complex Portfolio Directors
New Comps		Leadership Engagement Self Control Assertiveness Relaxation Openness Creativity Result Oriented Efficiency Consultation Negotiation Conflict Reliability Values Ethics	Project Team Member	Project Manager	Senior Project Manager	*Complicated* Project Director	Strategy Business Change Innovation OD Systems Leadership Culture Governance Attributes
		Program & Portfolio					
Traditional Comps		Integration Scope Time Cost Quality Risk HR Communication Procurement Safety					

Source: Dombkins, *Complex Project Management*, p. 300.

As shown in figure 10.5, the scope of competencies (certification levels A, B, C, and D) required by IPMA for traditional and complicated project managers has been expanded to now include a much broader range of behaviors and specific personal attributes. This is also reflected to a higher level in the CPI complex competency standards:

- IPMA integrates general management (GM) into its traditional (level B, C, and D) and complicated (level A) competency standards; and

- CPI integrates GM, ST (including systems engineering), system-of-systems, and complex program management, or CPM, into its complex competency standards.

GENERAL MANAGEMENT

General management, or GM, is based upon the machine metaphor and focuses on ongoing organizations. Organizational design, business process, long-range planning, and tools such as six-sigma are based upon stability and certainty. Over the past decade, the failure of strategic planning and the increased rate of environmental change have brought these assumptions of certainty under increasing pressure. In response, while still maintaining its philosophical foundations in certainty, GM has moved to stress the importance of leadership, emotional intelligence, empowerment, communication, alignment, and teams in providing flexibility and responsiveness. GM does not include ST.

SYSTEMS THINKING (INCLUDING SYSTEMS ENGINEERING)
As with GM, ST initially focused on certainty and the machine metaphor. However unlike GM, ST has developed a contingency approach that includes a continuum of approaches: at one node is systems engineering, based upon certainty and alignment in the environment, and at the other node are approaches based upon uncertainty and coercive environments.

COMPLEX PROGRAM MANAGEMENT
Unlike TPM, GM, and ST, which are founded on a philosophy based on certainty, CPM was specifically developed to be philosophically based upon uncertainty and chaos. Although CPM uses project management as an entry gateway, its competency framework, underpinning knowledge, and tools are built upon a broad range of other disciplines that deal with various aspects of complexity. Complex program management uses

- TPM to deliver short-term projects where there is scope certainty;
- wave planning and governance contracts to deliver programs with uncertain scope; and
- double-loop learning to periodically reframe the program.

COMPLEX COMPETENCY STANDARDS
The complex competency standards[8] are significantly different from TPM competency standards. The complex standards have nine views. The push-pull of the nine views of the complex standards provide a holistic understanding. The standards

- are based upon a dialectic of complexity and uncertainty and an emergence-based paradigm;
- use multiple views and dialectics to define behaviors that together provide insight and understanding;
- require a substantial level of underpinning knowledge; and
- require demonstrated special attributes.

Complex Paradigm
The standards are based on a complex program management paradigm. The complex paradigm is central to the complex standard and encompasses aspects including chaos, emergence, double-loop learning, and change.

Multiple Views
The nine views of the standard define behaviors and are based upon the premise that you cannot understand a whole through analyzing its parts (see figure 10.6). The nine views provide the following:

- insights from multiple perspectives, that together provide holistic understanding;
  - View 1. Strategy and program management
  - View 2. Business planning, life cycle management, reporting and performance measurement
  - View 3. Change and journey
  - View 4. Innovation, creativity, and working smarter
  - View 5. Organizational architecture
  - View 6. Systems thinking, SOS, and integration
  - View 7. Leadership
  - View 8. Culture and being human
  - View 9. Probity and governance

Figure 10.6. The Multiple Views of the Complex Competency Standards

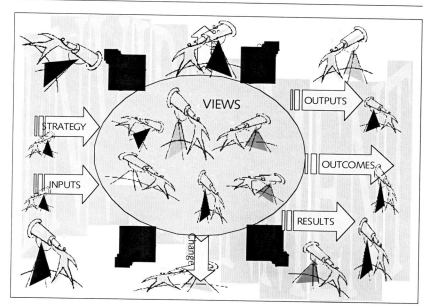

- a holistic understanding of the competencies required for the program management of complexity and the assessment of individuals against those competencies can only be achieved through using multiple views.

Underpinning Knowledge

Unlike TPM where underpinning knowledge is a minor factor in competency assessment, underpinning knowledge plays a significant role in the competency assessment of program directors, complex program directors, and complex portfolio directors (figure 10.7). They require a deep and broad underpinning knowledge:

- They must be competent across a broad range of areas and correspondingly require a significant breadth of knowledge. The complex competency standards draw their theoretical base from a broad range of literatures, and the depth of underpinning knowledge they require is moderated through four levels.
- Complex programs are "one-offs"—that is, they do not repeat themselves—with past projects at best providing insight only.

Figure 10.7. Role of Underpinning Knowledge in Competency Assessment

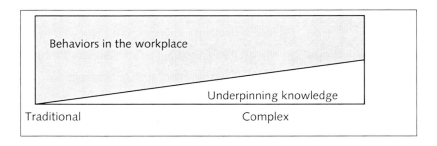

They require significant depth of knowledge in key areas to support the programs' architecture design.

- Given the breadth and depth of underpinning knowledge required, the most appropriate means for delivering this underpinning knowledge is through one-year full-time courses at tertiary business schools. Such courses must

  — layer the delivery of underpinning knowledge, first laying foundations in ST, strategy, change and journey management, and self-understanding and subsequently progressively building in the remaining views;

  — use experiential learning to build competencies in using the views; and

  — include robust mentoring and self-development.

Special Attributes
Although the paradigm defines the mindset and the views define the behaviors in the workplace, five special attributes define the personal characteristics essential for complex program and portfolio directors: wisdom and self-awareness, action/outcome orientation, ability to create and lead innovative teams, focus and courage, and ability to influence (see figure 10.8). The special attributes deliver two critical abilities:

- Provide leadership and a pathway forward when continuously confronted with multiple and opposing paradigms, views, and dialectics; and

Figure 10.8. Special Attributes

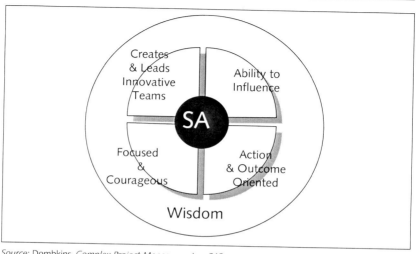

Source: Dombkins, *Complex Project Management*, p. 313.

- Enable individuals to not only survive, but flourish in what many would perceive as a high-pressure, personally demanding environment.

Each attribute provides a different focus and perspective. Holism is only achieved by looking at all the attributes.

1. Wisdom and Self-Awareness

Based upon research by Levinson[9] and the author's own professional experience and academic research, the categories of young and old are not tied to specific ages, but rather to psychological, biological, and social qualities. As individuals progress through their life cycle, they reach gateway points at which they make decisions that significantly shape their future. These gateway points represent periods of transition that modify the individual's life vision and put it into a new context.

Wisdom is developed by passing through these gateway points throughout the life cycle, during which the individual's psyche evolves from an undifferentiated image into an increasingly complex internal figure that maintains a dialectic of young and old. Although the internal young psyche maintains significant energy and capacity for further development, the internal old psyche has already reached its potential.

The location of these gateway points within the life cycle are commonly accepted among ancient scholars, including Confucius, Solon, and the writers of the Talmud.

A key characteristic is that the individuals do not believe they have yet reached their full potential. To stretch their personal potential, they continually explore and redefine the dialectic between young and old.

2. Action/outcome orientation

The drive to take action and the desire to deliver outcomes are essential. No matter what obstacles or resistance the individuals inevitably encounter, they remain focused on delivering the project outcomes.

3. Ability to create and lead innovative teams

Complex program and portfolio directors lead, inspire, and provide the energy to teams, enabling them to deliver more, both individually and synergistically, than they have ever previously achieved. They utilize their broad range of experience to drive creativity by providing the building blocks required to seed ideas within teams.

4. Focus and courage

Complex program and portfolio directors lead from the front and possess the courage to push boundaries and make hard decisions.

5. Ability to influence

A significant ability to influence others is essential for complex program and portfolio directors—in many instances, it is only through this special attribute that program support is achieved.

Certification by the CPI requires a candidate to satisfy four requirements:

- Proven competence in TPM and GM, the initial gateway for a complex program director, as most complex programs include subprojects that are PCAT 3, 4, or 5;
- Possession of the special attributes and an ability to work in both certainty and complexity/uncertainty-based paradigms;
- Proven underpinning knowledge for each of the nine views (representing distinct actions in the workplace); and
- Proven competence in each of the nine views.

Figure 10.9. Career Pathway and Training Strategy for Project Managers

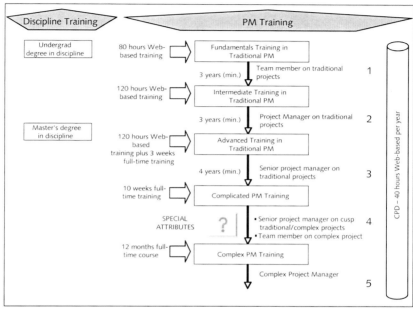

Source: Author's research and critical analysis of project management training and development for the U.S. DOD, Australian Defense Materiel Organization, or DMO, and Singapore Defense Science and Technology Agency, or DSTA; also, Dombkins, *Complex Project Management*, p. 316.

LIFE CYCLE DEVELOPMENT OF PROJECT MANAGERS

Project management currently lacks an internationally accepted integrated career pathway and training framework. Figure 10.9 proposes a generic career pathway and training strategy for project managers:[10]

The career pathway for project managers has five distinct phases:

1. **Team member:** person undertakes fundamentals training (skill-based) in project management and starts work as a project team member. During this period, the person is certified as a project team member.

2. **Project manager:** after a minimum of three years project management experience as a project team member, a person completes intermediate training and certification to become a project manager on traditional projects, or becomes a specialist scheduler, estimator, etc.

3. **Senior project manager on traditional projects:** after a further three years of experience as a project manager, the person completes

Figure 10.10. Certification Levels for Project Managers

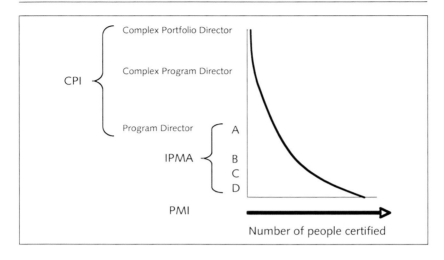

advanced training (skill-based) and is certified as a senior project manager for traditional projects.

4. **Complicated program director:** after an additional four years of experience as a senior project manager, the person completes training in complicated program management and is certified as a project director (IPMA Level A), and program director (CPI). Complicated program management training moves away from skill-based training to experiential learning focused on personal development and underpinning knowledge.

5. **Complex program or portfolio director:** after an additional five years experience as a complicated project/program director, the person undertakes a CPI-accredited course in complex program management. Upon completing a complex program management course under the mentorship of an existing CPI member, the person is assessed and then certified through peer review by the CPI.

The demand for project managers at each stage varies enormously: although there is a global shortage of certified project managers, this shortage is most pronounced for complicated and complex program directors. Strategies must be developed to fast-track the development of project managers to these certification levels (see figure 10.10).

EXISTING PROJECT MANAGEMENT CERTIFICATION FRAMEWORKS

Competency standards vary significantly in structure and scope. To be effective, all project management competency standards should

- be integrated with project types (PCAT);
- specifically state the paradigm in which it operates;
- provide adequate scope and depth;
- require an appropriate length and depth of training;
- specifically define depth and breadth of underpinning knowledge;
- be linked to maturity;
- require ongoing Continuous Professional Development (CPD);
- be supported by a robust, valid, and reliable assessment system;
- define required attributes; and
- provide a project management career pathway.

The current status of traditional, complicated, and complex certification requirements are as follows:

Traditional Projects

- The revised International Project Management Association (IPMA) competency standards integrate new competencies and attributes into certification requirements for traditional project managers. However, while the existing (IPMA) certification covers the required body of knowledge in project management, it lacks adequate depth and needs to be further revised to add significantly greater functional depth at each level and to provide specific requirements for underpinning knowledge;
- Existing project management certification frameworks do not provide for specialization in scheduling, estimating, risk, etc.; and
- Existing project management certification frameworks lack requirements for CPD.

Complicated Programs

- Complicated brings together the competencies required for TPM and many of the competencies of CPM (at varying levels of depth) at a greater depth in integration;
- The revised IPMA and the CPI Competency Standard[11] integrate new competencies and attributes into certification requirements;
- Assessment for complicated program directors requires peer review;
- Complicated program directors are required to work across both the TPM and CPM mindset; and
- The demand for complicated program directors is rapidly increasing in light of the escalated uncertainty and political factors in many programs.

Complex Programs and Portfolios

- Complex program and portfolio directors are best defined by the special attributes;
- Complex program and portfolio directors require significant breadth and depth in their underpinning knowledge: in most programs, experience alone is insufficient; and
- An approximate shortfall of 90 percent exists in the supply of complex program and portfolio directors.

PROPOSED MODEL FOR DEVELOPMENT OF PROJECT MANAGERS

Project managers' life cycle development must be strategically managed. The U.S. DOD and the Singapore Defense Science and Technology Agency (DSTA) provide case studies for the successful strategic management of the project management life cycle training and development of traditional and complicated project managers.[12]

Based upon observations of project manager training and development strategies in the Defense Departments of the United States, Australia, United Kingdom, and Singapore, the following key issues need to be considered in designing a strategy to develop project managers:

- Training must start by expressly defining the project manager's career pathway, the different paradigms that exist within project management, and the required competencies, experiential learning, and personal development;

- The overall quantum of training and education required must reflect that of other professions. The U.S. model provides a benchmark based on life cycle and maturity, although the strategy is heavily biased toward specific U.S. DOD systems skills training and lacks training in underpinning knowledge for complicated and complex levels;

- The timing for the delivery of project management training must be staged, to relate to experiential relevance and maturity; delivering high-level training to persons lacking sufficient maturity will have little benefit;

- The balance between behaviors in the workplace and underpinning knowledge changes throughout the project manager's life cycle—for example, for complex project managers there is a change in focus from behaviors in the workplace to underpinning knowledge;

- The depth and scope of underpinning knowledge also change throughout the project manager's life cycle;

- The delivery methodology for training/education must be fit-for-purpose and move from a Web-based training approach at entry to class-based methodologies during the intermediate and advanced phases and finally into group-based experiential education at the complicated and complex levels;

- The length, depth, and breadth of experience that are required to be certified as competent at each certification level;

- Development requires ongoing mentoring and coaching; and

- On-the-job based project management development strategies must be supported by a robust CPD program similar to that used by the U.S. DOD.

Table 10.4 proposes an integrated strategy for the life cycle training, education and development of project managers.

Table 10.4. Integrated Strategy for the Life Cycle Training, Education, and Development of Project Managers

Age	20–25	25–30	30–35	35–40	40–45	45–50	Total Training Hours	Training/Education Delivery Method
Discipline education (engineering, law, etc.)	Undergraduate		Graduate					
Project management training/education	Fundamentals	Intermediate	Advanced / Complicated	Complex Masters			60 80 300 400 1600	60 hrs. Web or equivalent 80 hrs. Web or equivalent 60 hrs. Web + 6 wks. residential 10 weeks residential 12 months full-time
Continuous professional development (CPD)	200 hrs.	200 hrs.	200 hrs.	200 hrs.	200 hrs.	200 hrs.	1200	40 hours per year
Total training/education hours (excluding discipline)							3640	Certification level
PM certification	Traditional Level 1	Traditional Level 2 Traditional Level 3	Complicated		Complex			IPMA level D IPMA level C IPMA level B IPMA level A CPI member/fellow

Source: Dombkins, Complex Project Management, p. 320.

FAST-TRACKING THE DEVELOPMENT OF COMPLEX PROGRAM AND PORTFOLIO DIRECTORS

To address the worldwide shortage in the supply of complex program and portfolio directors requires that we establish a methodology to identify high-potential candidates and to fast-track their development, given that

- little to no succession planning is in place;
- until recently, no suitable education programs have been available; and
- demand is growing.

Biographical, academic, and psychometric analysis of the CPI complex portfolio directors and complex program directors[13] presents a template to describe them:

- many did not fit well with traditional education programs;
- many honed their special attributes when they were in their mid-forties;
- they are not characterized by a common Myers-Briggs personality type;
- they have very high energy;
- they show great staying power in their personal relationships;
- they have mindsets that encompass both certainty and complexity/uncertainty;
- they are comfortable with ambiguity;
- they see each new project as a new challenge and their last project as their best;
- they constantly seek learning;
- they possess broad experience in both traditional and complicated project management;
- they possess a broad and deep underpinning knowledge; and
- they display the special attributes in their behaviors.

Using this template as a base, the key issue in attempting to fast-track the development of a young cohort of individuals is whether an old head can be put on young shoulders.

To date, the only reliable means of assessing the mindset and special attributes of potential complex program directors has been through the long-term assessment by a complex program director. Three factors must be taken into consideration in developing any methodology to identify high-potential candidates:

- academic testing is not a reliable indicator;
- psychometric testing, though theoretically appealing, remains in practice useful only in assisting individual development and/or when used in conjunction with assessment by a complex program director; and
- the complex mindset and the special attributes must underpin the methodology.

Once suitable candidates have been identified, a focused post-graduate course is required to deliver the required underpinning knowledge for complex program directors, to provide experiential learning in developing competencies within the nine CPI views, and to develop the special attributes within the individual. Although appropriately structured post-graduate courses are able to accelerate personal development, they cannot develop wisdom.

Wisdom is developed throughout an individual's life cycle. Wisdom can only be accelerated by using experiential learning (the "school of hard knocks") accompanied by mentoring to decrease the period between gateway points and thereby accelerate the psyche's evolution.

To develop high-potential candidates requires that both underpinning knowledge and self-understanding be provided through focused post-graduate courses and that wisdom be cultivated through experiential learning and mentoring.

KEY ELEMENTS IN A DEVELOPMENT PROGRAM FOR COMPLEX PROGRAM DIRECTORS

Across the board, existing training programs for the development of complex program directors are inadequate. Although some programs provide experiential learning utilizing case studies, to date no existing program has provided all of the following key elements:

- a structured process to select high-potential candidates who possess the core special attributes;
- a team-based learning environment where a small class (maximum 30) are cloistered for 12 months away in a skunk group–type environment;
- world class education in the breadth and depth in underpinning knowledge;
- a staged experiential learning experience to facilitate use of the views and their respective behaviors in the workplace;
- a structured mentoring program that extends for several years beyond the course;
- integrates personal development as a key aspect of the course;
- uses systems thinking and wave planning as a core for the course; and
- deliberately develops a complex paradigm (mindset) within the students.

Given the depth and breadth of knowledge to be imparted and the level of personal change required, it is unlikely that an effective complex program director development course could be any less than one year full-time. Given the scarce supply of high-quality candidates, the most effective courses for the development of complex program directors will be sandwich courses taken in blocks of study time and surrounding a period of practical experience.

PILOT EDUCATION PROGRAM

The Australian government has sponsored and funded a pilot executive course toward the master of business administration in complex project management.[14] The course is based on the CPI complex competency standards. Key features of the course are as follows:

- Selection of candidates—the development of a methodology to identify high-potential complex project manager candidates. The course is housed in a specially built facility where the students work as a skunk group, isolated from day-to-day operations.
- Key themes of systems thinking, wave planning, paradigm change, and personal development are threaded through all units.

- The course has 22 half-semester units plus 2 units composing an international study tour. Two of the units are work-integrated units.

- Each unit presents the key theories in a topic, reflecting the breadth and depth of the underpinning knowledge defined in the complex competency standards. The intent is to highlight and juxtapose opposing theories within each area and to understand their relevance from a systems view.

- The underpinning knowledge builds on a broad range of literatures that have been focused toward complex program management. The development of these high-level units that integrate leading theories, papers from leading international journals, complex program management, specially written case studies, and experiential learning provides a unique learning opportunity.

- The architecture of the course is structured so that a foundational knowledge is established in the first 6 units. The next 6 units require students to not only cover the unit-specific material, but just as importantly to concurrently use the views from the first six subjects. The same process is repeated for the next 6 units where students must concurrently use the views from the previous 12 units, etc., where students are concurrently using views from all 22 units.

- Case studies in complex program management have been specifically developed by organizations including Boeing, Lockheed Martin, and Raytheon.

- The selection of candidates and formal coursework are part of a development program that includes ongoing mentoring, and focused experiential learning through strategically planned job placement upon graduation.

CONCLUSION AND THE WAY FORWARD

Project management has developed from a vocation and is now a profession.[15] In so doing, it is integrating into itself underpinning knowledge from a broad range of disciplines and has established a robust philosophical base. As with other professions, project management is establishing an internationally recognized certification process that defines competencies, a career pathway, and certification for practitioner (traditional program manager) and specialist levels (complicated and complex program managers).

The CPI is the custodian of the Complex Program and Portfolio Director Competency Standards. It provides a reliable and valid standard upon which an individual's competencies, underpinning knowledge, and special attributes can be tested. The standards can also be used to assess an organization's competencies.

Through viewing projects as systems, it is obvious that, depending upon the complexity of the project/program, different competencies, implementation strategies, architectures, processes, tool sets, and contractual models are required. The development of the Project Categorization Framework, or PCAT, provides a methodology to categorize projects and programs according to their complexity and thereby defines competencies, implementation strategy, architecture, processes, tools, and contracts for each PCAT category.

Linking IPMA and CPI standards provides a structured career pathway. A project manager progresses along the career pathway through these steps:

- For a traditional project manager to become a complicated program director, a range of additional competencies are required and the importance of underpinning knowledge and attributes increases.

- For a complicated program director to become a complex program director requires the individual to

    — exhibit the ability to work effectively in two different paradigms simultaneously;

    — demonstrate possession of the special attributes;

    — show the mastery of complex specific competencies; and

    — possess substantial underpinning knowledge.

Currently a critical shortfall exists in the supply of competent complicated and complex program directors. This shortfall will worsen, given that the average age of complex portfolio and program directors in the CPI is 55 years of age. This shortfall in supply can only be overcome by fast-tracking the development of high-potential candidates in a hothouse environment.

Competent project managers are one part of a system in the successful development and implementation of complex programs. Complex programs require different governance, strategies, architectures, methodologies, tools, processes, and contracts.[16] Given the significant structural and cultural differences between a Traditional/Complicated Project Management Office (PMO) and a Complex Enterprise Program Management System (EPMS), organizations will need to establish EPMS as a separate unit.

Complex program management is purposefully designed to deliver highly complex programs and portfolios, such as those found in sustainability, health, environment, aid, infrastructure, defense, and technology. It is through successfully delivering these important and high-impact programs that complex program management will add significant value to our society.

NOTES

1. U.S. Department of Defense, Office of the Secretary of Defense (OSD), *Systems Engineering Guide for Systems of Systems*, version 1.0 (pre-release draft), June 2008, includes input from the author on emergence and systems architecture.

2. TPM is based upon project certainty—certainty in both project scope and project context (environment). TPM was initially based upon three outcomes—time, cost, and quality—with trade-offs being made among them. TPM's toolset has since expanded to now include nine tools with matching sets of competency standards.

3. The Complex Program Institute is the peak professional body for program directors and complex program and portfolio directors. The complex competency standard were developed with input from the Australian, U.S., and UK governments and multinational corporations including Boeing, Lockheed Martin, Raytheon, KBR, Thales, and BAE.

4. D.H. Dombkins, *Complex Project Management* (Charleston, S.C.: BookSurge, 2007).

5. D.H. Dombkins and P. Dombkins, *Contracts for Complex Programs* (Charleston, S.C.: BookSurge, 2008).

6. Traditional positivist approaches to ST (including systems engineering) focus upon facts, while complex ST systems approaches use multiple views and dialectics.

7. DOD, *Systems Engineering Guide for Systems of Systems*, version 1.0 (pre-release draft), June 2008.

8. David H. Dombkins authored the Complex Program and Portfolio Director Competency Standards, available at http://www.complexpm.com/.

9. Daniel J. Levinson, *The Seasons of a Man's Life* (New York: Knopf, 1978), p. 34: "Greater wisdom regarding the external world can only be gained through a strong centering of the self; and some of the greatest intellectual and artistic works have been produced by men in their sixties, seventies and even eighties."

10. Based upon the author's research and analysis of project management training and development for the U.S. DOD, Australian Defense Materiel Organization, or DMO, and Singapore Defense Science and Technology Agency, or DSTA.

11. The CPI complex competency standards define competencies for both complicated and complex.

12. See Dombkins, *Complex Project Management,* pp. 289–345, for analysis of the life cycle development strategies used by the U.S. DOD, the Australian DMO, and the Singapore DSTA.

13. Twelve of the world's leading complex portfolio and program directors.

14. The author has conceived, designed, and supervised the development of the course.

15. D.H. Dombkins, "Redefining My Profession—Part 2," IPMA World Congress, Rome, 2008.

16. Dombkins and Dombkins, *Contracts for Complex Programs.*

ABOUT THE EDITORS AND AUTHORS

Julie M. Anderson leads a variety of strategic policy initiatives for IBM's global public sector business and manages the Global Leadership Initiative. Prior to joining IBM, Ms. Anderson served as a legislative aide to Senator J. Robert Kerrey of Nebraska and worked as a policy analyst in the Office of the U.S. Secretary of Transportation. She is a Harry S. Truman Scholar. Ms. Anderson is coauthor of *Living Well: Transforming America's Health Care* (IBM Global Services, 2008). She earned an M.B.A. from Duke University, an M.P.P. from the University of Chicago, and a B.S. from Nebraska Wesleyan University.

Guy Ben-Ari is a fellow with the Defense-Industrial Initiatives Group at CSIS, where he specializes in defense technology and defense industrial policies. Before joining CSIS, he was a research associate at the George Washington University's Center for International Science and Technology Policy as well as a consultant focusing on innovation policy and evaluation for the European Commission and the World Bank. He is coauthor (with Gordon Adams) of *Transforming European Militaries: Coalition Operations and the Technology Gap* (Routledge, 2006) and of various book chapters and articles. He holds a master's degree in science, technology, and public policy from the George Washington University and a bachelor's degree in political science and history from Tel Aviv University.

David J. Berteau is senior adviser and director of the CSIS Defense-Industrial Initiatives Group. A former director of Syracuse University's

National Security Studies Program, Mr. Berteau is an adjunct professor at Georgetown University, a member of the Defense Acquisition University Board of Visitors, and a director of the Procurement Round Table. He is a fellow of the National Academy of Public Administration and a member of the Federal Outreach Advisory Committee of the Association of Defense Communities. Prior to joining CSIS, he was director of national defense and homeland security for Clark & Weinstock, a senior vice president at Science Applications International Corporation (SAIC), and principal deputy assistant secretary of defense for production and logistics. Mr. Berteau holds a B.A. from Tulane University and a master's degree from the LBJ School of Public Affairs at the University of Texas.

Pierre A. Chao is a senior associate with the Defense-Industrial Initiatives Group at CSIS. Between 2003 and 2007, he was a senior fellow at CSIS. Prior to joining CSIS, he spent eleven years on Wall Street, the last four as a managing director at Credit Suisse First Boston. Mr. Chao served on the Presidential Commission on Offsets in International Trade and on seven Defense Science Board and Defense Business Board studies, and he is a member of the Aeronautics and Space Engineering Board of the National Research Council. He earned dual B.S. degrees from MIT.

David H. Dombkins is a complex program director, consultant, and academic with more than 35 years experience. He has developed innovative strategic solutions for some of the world's most uniquely complex and pioneering projects in change, defense, ICT, mining, energy, infrastructure and outsourcing. He pioneered the development of relational contracting, developed the Complex Project Managers Competency Standard, and authored the books *Complex Project Management* (BookSurge, 2007) and *Contracts for Complex Programs* (BookSurge, 2008). He holds a doctorate of technology degree from Deakin University.

Jeffrey A. Drezner is a senior policy researcher based in RAND's Santa Monica office. He has conducted policy analysis on a wide range of issues, including energy research and development; planning and program management; best practices in environmental management; analyses of cost and schedule outcomes in complex system development programs; aerospace industrial policy; defense acquisition policy and reform; and local emergency response. He holds an M.S. in policy analysis from the University of California, Davis, and a Ph.D. in political science from the Claremont Graduate School.

Eugene Gholz is an associate professor at the LBJ School of Public Affairs at the University of Texas in Austin. His research focuses on innovation, business-government relations, defense management, and U.S. foreign military policy. He is also a research affiliate of MIT's Security Studies Program and associate editor of the journal *Security Studies*. Before moving to Texas, he was assistant director and assistant professor at the University of Kentucky's Patterson School of Diplomacy and International Commerce. He is the coauthor (with Peter Dombrowski) of *Buying Military Transformation: Technological Innovation and the Defense Industry* (Columbia University Press, 2006) and (with Harvey M. Sapolsky and Caitlin Talmadge) of *U.S. Defense Politics: The Origin of Security Policies* (Routledge, 2008). He received a Ph.D. from MIT.

W. Scott Gould directs strategy formulation for IBM's homeland security, intelligence, and defense business and directs IBM's Global Leadership Initiative. Previously, he was CEO of The O'Gara Company, providing strategic advisory and investment services in the homeland security market. He has also served as CFO and assistant secretary for administration at the Commerce Department, as deputy assistant secretary for finance and management at the Treasury Department, and as a White House fellow. He is coauthor of *The People Factor: Strengthening America by Investing in the Public Service* (Brookings Institution, forthcoming). He holds an A.B. degree from Cornell University and M.B.A. and Ed.D. degrees from the University of Rochester.

Marco Iansiti, originally a physicist, joined the faculty of the Harvard Business School in 1989. His research has focused on technological innovation, product development, entrepreneurship, and operations seeking the drivers of productivity, flexibility, and adaptation in organizations. He currently investigates the implementation of successful strategies for companies that wish to better understand and manage innovation and operations in complex networked industries, also known as "business ecosystems." Professor Iansiti has authored and coauthored more than 50 articles, papers, book chapters, cases, and notes, including *Technology Integration: Making Critical Choices in a Dynamic World* (Harvard Business School Press, 1997) and *The Keystone Advantage: What the New Dynamics of Business Ecosystems Mean for Strategy, Innovation, and Sustainability* (Harvard Business School Press, 2004).

Jeremy M. Kaplan is a professor and holds the Defense Information Systems Agency (DISA) Chair at the Industrial College of the Armed Forces, National Defense University. He was formerly the director for Technical Integration Services in DISA, the director for Modeling and Simulation in DISA, deputy director of the C4I Integration Support Activity in ASD/C3I, the DISA/JIEO director of the Center for (DOD) Information Technology Standards, and the DISA/C3S deputy director for Strategic C3.

Douglas O. Norman is the director for complex systems engineering in MITRE's Command and Control Center, a small group responsible for cross-corporate systems engineering at enterprise scales. Previously, Mr. Norman was chief engineer for Theater Battle Management Core Systems (TBMCS), the key system used in Operation Iraqi Freedom to plan and manage the air war, and for Air and Space Operations Center—Weapon System (AOC). Additionally, he is interested in the business of technology and has both started businesses and mentored business and technology students who have gone on to start high-tech businesses.

Harvey M. Sapolsky is professor of public policy and organization at MIT and the former director of the MIT Security Studies Program. He has published extensively on defense, health, and science policy issues and has served on various related government policy panels, committees, and commissions.

Michael Schrage is a senior adviser to MIT's Security Studies Program and research associate with the MIT Sloan School's Center for Digital Business. He also serves on the technical advisory committee for MIT's Lincoln Labs. He has worked with the Office of Net Assessment, the State Department, and the National Security Council on innovation-related issues. He helped launch Accenture's "Digital Governance" initiative, exploring how technology can improve oversight, accountability, and risk management.

Matthew Zlatnik is a research consultant with the Defense-Industrial Initiatives Group at CSIS, focusing on how technological, industrial, and budgetary issues affect defense policy. He previously spent 10 years in investment banking. Mr. Zlatnik graduated from Carleton College, holds an M.B.A. in finance from the Wharton School, and is studying for a master's degree in international relations at the Johns Hopkins University School of Advanced International Studies (SAIS).

INDEX

Page numbers followed by f *and* t *refer to information in figures and tables. Page numbers followed by the letter* n *refer to end-of-chapter notes.*

Academics
  as stakeholders, 120
  systems integration and, 52
Accountability, megaprojects and, 101
ACE. See Advanced Collaborative Environment (ACE)
Acquisitions, weapon
  arsenal model of, 25, 26t, 27
  community of interest in, 24
  competition and, 36, 40–42
  complexity in, 32–36
  complexity metrics and, 45
  contract model for, 26t, 27
  engineering and, 82
  environment for, 35
  evolutionary, 46
  innovation in, 40–42
  Lead Systems Integrator model for, 26t, 29
  outsourcing model for, 26t, 28–29
  political will and, 24
  value and, 83
  weapon system manager model for, 26t, 27–28
ACTDs. See Advanced Concept and Technology Demonstrations (ACTDs)
Action orientation, 144
Adaptability
  in IT-intensive systems, 82–83
  in systems integration, 54–55

Advanced Collaborative Environment (ACE), 104
Advanced Concept and Technology Demonstrations (ACTDs), 46
Advanced Technology Demonstrations (ATDs), 46
Advertising, 126
Advisory board, 123–124
Advocacy organizations
  environmental, 124
  as stakeholders, 120
Aerospace Corporation, 57
Alberts, David, 2
Anderson, Julie M., 7–8
Archetypes, project, 135t–136t
Arsenal model, of acquisition, 25, 26t, 27
Assumptions
  about innovation, 39–40
  reevaluation of, 2–3
  risk, 18
ATDs. See Advanced Technology Demonstrations (ATDs)
Australian pilot education program, 153–154
Authority
  in megaprojects, 100–101
  system-of-systems, 72–75, 74f
Awareness, technical, systems integration and, 51–52, 64t

Barriers
  entry, 36, 41
  organizational, 100
Bath Iron Works, 37, 47n7
Bear Stearns, 16
Big Dig (Boston), 90t
Bissell, Richard, 11
Board of directors
  CEO accountability and, 14
  expectations and, 14–15
Boeing, 41
Boeing's Advanced Collaborative Environment, 104
Bookstaber, Rick, 18
Boundaries, military, 68
Bracken, Paul, 38
Budgets, FCS and, 93f
Buffett, Warren, 14
Burden shifts, 114
Bureaucracy, innovation and, 44
Burke, Arleigh, 11
Bush, Vannevar, 21, 21n3
Business ecosystems, 96f

Campaign management, 123
Campaign manager role, 123–124
Capability
  creation, in enterprise development, 67, 121–123, 122f
  workforce, 42
Capability Maturity Models, 55
Career pathway, project manager, 145f
Case study, IBM, 125
CDOs. *See* Collateralized debt obligations (CDOs)
CEO. *See* Chief executive officer (CEO)
CEO Council, 104
Certainty
  paradigm of, 129, 131t
  scope, 134f
Certifications, project management, 131, 147–148
Change management, 117
Channel tunnel, 90t
Chaos, 131t
Characteristics, enterprise, 68–71, 70f
Chief executive officer (CEO), 14

Citigroup, 13, 16, 22n11
Classification, project, 130–131, 131f
Climate change, 124
Code, in FCS, 92
Collaborative environment, 76–79, 77f
Collateralized debt obligations (CDOs), 15–18
Commercial complexity, 132–133
Community of interest, 24
Competencies, program director, 134, 138–139, 139f
Competency standards, 139–140
Competition
  acquisition and, 36, 40–42
  complexity and, 40–42
  entry barriers and, 36, 41
  FFRDCs and, 62
  globalization and, 43
  history of, 47n6
  innovation and, 3, 37
  in mature sectors, 43
  microeconomics and, 36–37
  new sectors and, 43
  organizational complexity and, 43–44
  policy and, 43–46
  policy levers and, 45–46
  traditional views of, 36–40
Completeness, requirements and, 84
Complex competency standards, 139–140
Complexity
  acquisitions and, 40–42
  advantages of, 41–42
  assumptions and, 2–3
  breaking down, 1–2
  commercial, 132–133
  competition and, 40–42
  vs. complicatedness, 1, 128–129
  cost, 133
  definition of, 32–36
  disadvantages of, 41–42
  in DOD governance, 68–69
  dynamic, 128
  elements within, 2
  environmental, 32, 35, 42
  external system, 133
  increasing, 33–34

innovation and, 3
internal system, 132–133
of issues, increasing, 111
life cycle, 133
management, 45
in megaprojects, 89–90
metrics for, 45
as obstacle, 4–5
as opportunity, 5
optimization and, 67
organizational, 32, 34–35, 41, 42, 43–44
paradigm of, 129
pitfalls in research on, 5
of project management, 53
rankings, PCAT, 132*t*
risk and, 15
schedule, 133
scope, 134*f*
stability and, 83
stakeholder, 133
subsystems and, 33–34
team, 132–133
technical, 32, 33, 41, 47*n*4
technology and, 1
understanding of, 2
in weapons acquisition, 32–36
of weapons systems, 32
weapon system manager model and, 27–28
Complex Program Institute (CPI), 130, 131, 155, 156*n*3
Complex program management, 139
Complex project management (CPM), 129, 135*t*–136*t*
Complex systems thinking, 133–134, 134*f*
Complicatedness, *vs.* complexity, 1, 128–129
Complicated project management, 129–130, 135*t*–136*t*
Conant, James, 21
Conflicts of interest, systems integration and, 56, 61
Contextual design, collaborative engineering and, 77*f*
Contingency management, 120
Contracting

FFRDCs and, 62
innovation and, 46
systems integration and, 61
Contract model, for acquisition, 26*t*, 27
Contractors
increase in, 111
as policy partners, 118–119
as stakeholders, 112
Control trade-offs, 116
Coordinated action plan, 120
Coordinated sales process, 118
Coordination costs, 116
Corporate governance, 12–18
Cost complexity, 133
Cost-performance-agility trade-offs, 78
Counterinsurgency, 42
Courage, 144
CPI. *See* Complex Program Institute (CPI)
Culture, in enterprise development, 79–80
Currie, Malcolm, 11
Customers, in systems integration, 53–54, 64*t*
Customer service, 114
Cyber warfare, 42

DAPA. *See* Defense Acquisition Performance Assessment (DAPA) project
DARPA. *See* Defense Advanced Research Projects Agency (DARPA)
Data repositories, 77
DDG-1000. *See* Zumwalt class destroyers
Decomposition, of complexity, 1–2
Deepwater, 34
Defense Acquisition Performance Assessment (DAPA) project, 4
Defense Acquisition Transformation Report to Congress, 22*n*7
Defense Advanced Research Projects Agency (DARPA), 38
Defense Information Systems Agency (DISA), 70
Defense Information Systems Network (DISN), 71

Defense Science and Technology Agency (DSTA) (Singapore), 148
Defense Science Board Task Force on Developmental Test and Evaluation (report), 19
Deming, W. Edward, 75
Denver airport, 90*t*
Department of Defense (DOD). *See also* Government
　governance complexity, 68–69
　organizational complexity and, 41
　systems-of-systems in, 70–71
Design, process, 95, 97, 97*f*
Development, director, 151–154
Development, enterprise
　capability creation and, 67
　collaborative environment and, 76–79, 77*f*
　conceptual framework, 70*f*
　cost-performance-agility trade-offs in, 78
　culture and, 79–80
　developmental friction and, 69–70, 75
　developmental processes and, 72
　DOD governance and, 68–69
　enabling concepts, 70*f*
　enterprise characteristics and, 68–71, 70*f*
　enterprise size and, 69
　feature specification and, 67
　fundamental challenges, 66–68
　information flow and, 69
　initiative encouragement in, 75–76
　interoperability and, 67, 79
　net-centric guiding principles, 70*f*
　net-centric system-of-systems engineering, 71–76, 74*f*
　new technology and, 66
　posting requirements, 78
　problems in, 68–71, 70*f*
　simultaneous, 66
　single missions in, 68
　system-of-systems engineer in, 73–74, 74*f*
　system-of-systems authority and, 72–75, 74*f*
　systems-of-systems in, 68
　unity of purpose and, 75
　visibility and, 77–78, 77*f*
Development, of project managers, 148–149, 150*t*
Developmental friction, 69–70, 75
Dietrick, Robert A., 33
Difficulty, technical, 132
Direction, technical
　in arsenal model, 26*t*
　in contract model, 26*t*
　definition of, 25
　in Lead Systems Integrator model, 26*t*
　in outsourcing model, 26*t*
　in weapon system manager model, 26*t*
Director competencies, 134, 138–139, 139*f*
Director development, 151–154
Director level, 146
DISA. *See* Defense Information Systems Agency (DISA)
Discovery phase, 137*f*
DISN. *See* Defense Information Systems Network (DISN)
Distributed nature
　of FCS, 94*f*
　of megaprojects, 90
Diverse supplier networks, 60
Diversity, program, 46
DOD. *See* Department of Defense (DOD)
Dombkins, David H., 8
Double-loop learning, 129, 137*f*
Dowding, Hugh, 10
Drezner, Jeffrey A., 6
DSTA. *See* Singapore Defense Science and Technology Agency (DSTA)
Dubai Ports World, 121
Dynamic complexity, 128

Ecosystem efficiency, 102–103
Ecosystem management, 95, 96*f*, 108*t*
Ecosystem responsiveness, 103–104
Efficiency
　ecosystem, 102–103
　increasing demand for governmental, 113
Electric Boat, 37, 47*n*7

Emergence, 131*t*
End-user requirements, 110*n*16
Engineering
  acquisitions and, 82
  in isolationist approach, 85
  of materiel solutions, 84
  as multifaceted, 82
  operational value and, 86
  stability and, 84
Enron, 13
Enterprise characteristics, 68–71, 70*f*
Enterprise development
  capability creation and, 67
  collaborative environment and, 76–79, 77*f*
  conceptual framework, 70*f*
  cost-performance-agility trade-offs in, 78
  culture and, 79–80
  developmental friction and, 69–70, 75
  developmental processes and, 72
  DOD governance and, 68–69
  enabling concepts, 70*f*
  enterprise characteristics and, 68–71, 70*f*
  enterprise size and, 69
  feature specification and, 67
  fundamental challenges, 66–68
  information flow and, 69
  initiative encouragement in, 75–76
  interoperability and, 67, 79
  net-centric guiding principles, 70*f*
  net-centric system-of-systems engineering, 71–76, 74*f*
  new technology and, 66
  posting requirements, 78
  problems in, 68–71, 70*f*
  simultaneous, 66
  single missions in, 68
  system-of-systems engineers in, 73–74, 74*f*
  system-of-systems authority and, 72–75, 74*f*
  systems-of-systems in, 68
  unity of purpose and, 75
  visibility and, 77–78, 77*f*
Enterprise size, 69

Enterprise-wide system-of-systems engineering, 71–76, 74*f*
Entry barriers, 36, 41
Environment, external, 135*t*
  in arsenal model, 26*t*
  in contract model, 26*t*
  in Lead Systems Integrator model, 26*t*
  in outsourcing model, 26*t*
  in weapon system manager model, 26*t*
Environmental advocacy, 124
Environmental assessment, 122
Environmental complexity, 32, 35, 42
Environmental evolution, 124
Environmental Protection Agency (EPA), 124
EPA. *See* Environmental Protection Agency (EPA)
Evolution, of environment, 124
Evolutionary acquisition strategies, 46
Execution
  in arsenal model, 26*t*
  in contract model, 26*t*
  definition of, 25
  in Lead Systems Integrator model, 26*t*
  in outsourcing model, 26*t*
  support, 122
  in weapon system manager model, 26*t*
Executive sponsor role, 123
Expectations, boards of directors and, 14–15
Experience
  megaprojects and, 105–106
  systems integration and, 56
Experimentation
  innovation and, 106
  megaprojects and, 105–106
Expertise, manufacturing, 64*t*
External environment, 135*t*
  in arsenal model, 26*t*
  in contract model, 26*t*
  in Lead Systems Integrator model, 26*t*
  in outsourcing model, 26*t*
  in weapon system manager model, 26*t*
External stakeholders, 112
External system complexity, 133

Fast-tracking, of director development, 151–154
FCS. *See* Future Combat System (FCS)
Feature specification, 67
Federally funded research and development centers (FFRDCs), 57, 62
Feedback, 123
Feedback loops, nonlinear, 128
FFRDCs. *See* Federally funded research and development centers (FFRDCs)
Financial crisis, 15–18
Financial engineers, 16
Flexibility
  complexity and, 2
  innovation and, 40
  in private sector, 3–4
Focus, 144
Foresight, governance and, 20
Frequency, program, 46
Friction, developmental, 69–70, 75
Functional Capabilities Boards, 68, 72
Future Combat System (FCS), 33, 34, 35, 88–89, 91*f*, 92–93, 93*f*, 94*f*. *See also* Megaprojects
Future Force, 88, 108*n*2

GCCS. *See* Global Command and Control System (GCCS)
GCSS. *See* Global Combat Support System (GCSS)
General Dynamics, 47*n*7
General management (GM), 138–144, 141*f*, 142*f*, 143*f*
Gholz, Eugene, 6
GIG. *See* Global Information Grid (GIG)
Global Combat Support System (GCSS), 71
Global Command and Control System (GCCS), 70–71
Global financial crisis, 15–18
Global Information Grid (GIG), 71, 76–81, 77*f*
Globalization, 43
Global Movement Management (GMM), 125
Global positioning system (GPS), 11
Global warming, 124
GM. *See* General management (GM)
GMM. *See* Global Movement Management (GMM)
Goldman Sachs, 17, 22*n*11
Goldwater-Nichols command reorganization, 20
Gould, W. Scott, 7–8
Governance
  assumptions and, 18
  as challenge to leadership, 19
  corporate, 12–18
  DOD, 68–69
  expectations of, 16
  foresight and, 20
  insight and, 20
  vs. leadership, 11–12
  multiple authorities, 69
  overlapping, 69
  oversight and, 20
  reform, 14
  risk and, 15
  testing and, 19
  as underappreciated factor, 11
Government. *See also* Department of Defense (DOD)
  in arsenal model, 26*t*
  in contract model, 26*t*
  efficiency in, demand for, 113
  engagement of, 113
  environment, complexity of, 115–116
  incompatible goals in, 4
  as network, 111–112, 112–114
  in outsourcing model, 26*t*
  performance pressure on, 114
  private sector collaboration, to manage complexity, 116–117
  problems, complexity of, 115–116
  as stakeholder, 112
  stakeholders in, 112
  in weapon system manager model, 26*t*
Government database, 113
Government laboratories, 58–59, 64*t*
Government relations firms, as stakeholders, 120

Government services
  IBM case study in, 125
  industry expansion, 116
GPS. *See* Global positioning system (GPS)
Groves, Leslie, 21

HAE-UAV. *See* High-altitude endurance unmanned aerial vehicle (HAE-UAV)
Hardin, Garrett, 2
Hayes, Richard, 2
High-altitude endurance unmanned aerial vehicle (HAE-UAV), 45

Iansiti, Marco, 7
IBM, 123, 125
ICBMs. *See* Intercontinental ballistic missiles (ICBMs)
Identification, of stakeholders, 120
Incremental build, 137*f*
Independent systems-of-systems, 70
Industry
  arsenal model and, 27
  conflicts of interest and, 61
  in contract model, 26*t*, 27
  expansion of, 116
  flexibility in, 3–4
  goals of, *vs.* government, 4
  increased use of, 119
  innovation in, *vs.* public sector, 40
  in Lead Systems Integrator model, 26*t*
  organizational complexity and, 34, 41
  in outsourcing model, 26*t*, 28–29
  as policy partner, 118–119
  as stakeholder, 112
  systems integration by, 59–61
  technical complexity and, 41
  viability of base, 36
  in weapon system manager model, 26*t*, 28
Influence generation, 122
Influencer identification, 120
Information flow
  in enterprise development, 69
  through stakeholders, 113
  Web-enabled, 113
In-house laboratories, 58–59, 64*t*

Innovation
  assumptions about, 39–40
  bureaucracy and, 44
  competition and, 37
  complexity and, 3
  conditions and, 38
  contracting and, 46
  entry barriers and, 41
  environmental complexity and, 42
  expected results, 37
  experimentation and, 106
  factors, 38–39
  financial, 16
  firm size and, 40
  flexibility and, 40
  institutional environment and, 39
  investment and, 38
  in megaprojects, 89
  national factors, 38
  niche firms and, 40
  organizational complexity and, 42
  origins of, 38
  policy levers and, 45–46
  in private *vs.* public sector, 40
  regulatory environment and, 39
  status and, 38
  supporting industries and, 38
  technical complexity and, 42
  threat nature and, 42
  traditional views of, 36–40
  in weapons acquisition, 40–42
Insight, governance and, 20
Institutional environment, innovation and, 39
Integration, process, 104–105
Integration, systems
  academics and, 52
  adaptability in, 54–55
  conflicts of interest and, 56, 61
  contracting and, 61
  customer and, 53–54, 64*t*
  definition of, 50
  demand for, 51
  experience and, 56
  federally funded research and development centers in, 57, 62–63
  hallmarks of ideal, 51–57

Integration, systems *(continued)*
  increasing importance of, 64–65
  manufacturing expertise and, 64t
  organizational independence and, 59, 64t
  organizational longevity and, 64t
  organization types for, 57–64, 64t
  by prime contractors, 59–61
  project management and, 52–53, 64t
  skill in, 51
  technical awareness and, 51–52, 64t
  transparency in, 56
  trust and, 59
Integrity, process, 95, 97, 97f, 104–106, 108t
Intercontinental ballistic missiles (ICBMs), 11
Internal system complexity, 132–133
International Project Management Association (IPMA), 130, 131, 147, 155
Interoperability, 67, 79
Interorganizational relationships, 54
IPMA. *See* International Project Management Association (IPMA)
Iridium, 90t
Isolationist approach, 85
Issues, increasing complexity of, 111

JCIDS. *See* Joint Capabilities Integration and Development System (JCIDS)
JDAM. *See* Joint Direct Attack Munition (JDAM)
Jervis, Robert, 2
Johnson, Kelly, 11
Joint Capabilities Integration and Development System (JCIDS), 68
Joint Direct Attack Munition (JDAM), 45, 118–119
Joint Strike Fighter (JSF), 33, 34
JSF. *See* Joint Strike Fighter (JSF)

Kaplan, Jeremy M., 6–7
Key influencer identification, 120
Keystone organizations, 95
Knowledge, underpinning, 141–142

Laboratories, government, 64t
Leadership
  vs. governance, 11–12
  governance as challenge to, 19
  importance of, 11
  technical, 15
  termination of poor, 14
Lead Systems Integrator model, for acquisition, 26t, 29
Lead Systems Integrator (LSI), 29, 88, 98–106, 99f, 104
Learning, double-loop, 129, 137f
Levers, policy, 45–46
Life cycle archetype, 136t
Life cycle complexity, 133
Life cycle development, of project managers, 145–149, 145f, 146f, 150t
Life cycle phases, 133–134, 134f
Line replaceable modules (LRMs), 104
Lines of code, in FCS, 92
Lockheed Martin, 41
Longevity
  organizational, 64t
  of systems, 54–55
  of systems-of-systems, 69
Lorell, Mark, 34
LRMs. *See* Line replaceable modules (LRMs)
LSI. *See* Lead Systems Integrator (LSI)

MacLean, Bill, 11
Management, campaign, 123
Management, change, 117
Management, contingency, 120
Management, ecosystem, 95, 96f, 108t
Management, program
  in arsenal model, 26t
  authority, 100–101
  capability, 108f
  complex, 139
  complexity and, 45
  complicated, 129–130
  in contract model, 26t
  definition of, 25
  hierarchical structure in, 109n13
  keystone organizations in, 95
  in Lead Systems Integrator model, 26t

life cycle phases and, 133–134, 134*f*
lightweight, 96*f*
modes of, 95
organizational complexity and, 34
outcome scores, 108*t*
in outsourcing model, 26*t*
premier, 96*f*
theoretical frameworks, 98*f*
in weapon system manager model, 26*t*
Management, project
  accountability in, 101
  authority and, 100–101
  certifications, 131, 147–148
  challenges, 89–93, 90*t*, 91*f*, 93*f*
  changes and, 89
  classification in, 130–131, 131*f*
  complex, 129, 135*t*–136*t*
  complexity in, 89–90
  complexity of, 53
  complicated, 129–130, 135*t*–136*t*
  as continuum, 129–130
  distributed nature of, 90
  ecosystem efficiency and, 102–103
  ecosystem management and, 95, 96*f*, 108*t*
  ecosystem responsiveness and, 103–104
  experience and, 105–106
  experimentation and, 105–106
  FCS as, 92–93, 93*f*
  future of, 154–156
  life cycle phases and, 133–134, 134*f*
  managerial capabilities and, 93–98
  manager types in, 130
  nodes in, 129–130
  organizational barriers and, 100
  organization in, 99–100
  performance on management capabilities, 99*f*
  performance outcomes in selected, 90*t*
  pitfalls, 89–93, 90*t*, 91*f*, 93*f*
  process integration and, 104–105
  process integrity and, 104–106, 108*t*
  spectrum of, 129–130
  stakeholders in, 90
  structure, 99–100
  systems integration and, 52–53, 64*t*
  traditional, 129–130, 135*t*–136*t*, 156*n*2
  typology of projects in, 130–133, 131*f*, 131*t*, 132*t*
  uncertainty in, 90
Manhattan Project, 21
Manufacturing expertise, 64*t*
Marketing, 120, 126
McKinsey & Company, 3
Media
  as stakeholder, 120
  use of, 126
Megaprojects
  accountability in, 101
  authority in, 100–101
  capacity, 108*t*
  challenges, 89–93, 90*t*, 91*f*, 93*f*
  changes and, 89
  complexity in, 89–90
  distributed nature of, 90
  ecosystem efficiency and, 102–103
  ecosystem management and, 95, 96*f*, 108*t*
  ecosystem responsiveness and, 103–104
  experience and, 105–106
  experimentation and, 105–106
  FCS as, 92–93, 93*f*
  managerial capabilities and, 93–98
  organizational barriers and, 100
  organization in, 99–100
  outcome scores, 108*f*
  performance on management capabilities, 99*f*
  performance outcomes in selected, 90*t*
  pitfalls, 89–93, 90*t*, 91*f*, 93*f*
  process integration and, 104–105
  process integrity and, 104–106, 108*t*
  stakeholders in, 90
  structure, 99–100
  uncertainty in, 90
Memorandum of agreement (MOA), 85
Memorandum of understanding (MOU), 85
Merrill Lynch, 16, 22*n*11

Microeconomics, 36–37
Military service boundaries, 68
Millstein, Ira, 14
MITRE, 57
MOA. *See* Memorandum of agreement (MOA)
Modeling
  collaborative engineering and, 77*f*
  to manage complexity, 117
Modular build, 137*f*
Morgan Stanley, 18, 22*n*11
Mortgage meltdown, 15–18
MOU. *See* Memorandum of understanding (MOU)
Multiple perspectives, 140

National Defense Research Committee, 21*n*3
Net-centric guiding principles, 70*f*
Net-centric system-of-systems engineering, 71–76, 74*f*
NETWARS C3 model, 78–79
Network, government as, 111–112, 112–114
NGOs. *See* Nongovernmental organizations (NGOs)
Nongovernmental organizations (NGOs), as stakeholders, 120
Nonlinear feedback loops, 128
Nonprofit organizations, as stakeholders, 120
Norman, Douglas O., 7
Northrop Grumman Ship Systems, 37, 41, 47*n*7

Office of Scientific Research and Development (OSRD), 21*n*3
Operational value, 83–84, 86
Opportunity creation, 122
Optimization, complexity and, 67
Organization
  in megaprojects, 99–100
  of stakeholders, 113
Organizational complexity, 32, 34–35, 41, 42, 43–44
Organizational independence, systems integration and, 59, 64*t*
Organizational longevity, 64*t*

OSRD. *See* Office of Scientific Research and Development (OSRD)
OTA. *See* "Other transaction authority" (OTA)
"Other transaction authority" (OTA), 40
Outcome orientation, 144
Outsourcing model, for acquisition, 26*t*, 28–29
Outsourcing oversight, 113–114
Overlapping governance, 69
Oversight
  governance and, 20
  of outsourcing, 113–114

Packard Commission, 20
Parkinson, Brad, 11
PCAT. *See* Project Categorization Framework (PCAT)
PCAT numbers, 132*t*
PEOs. *See* Program executive officers (PEOs)
Performance pressure, 114
Permission sets, 119–120
Persuasion, of stakeholders, 117
PERT. *See* Program evaluation review technique (PERT)
Pierce, John R., 20
Pilot education program, Australian, 153–154
Pluralist environments, 129, 131*t*
Polaris, 11
Policy
  competition and, 43–46
  contractors as policy in, 118–119
  stakeholders and, 115
Policy levers, 45–46
Political will, 24
Politics, FCS and, 93*f*
Portfolio approach, 126
Portfolio director, 146
Portfolio director development, 151–154
Port operations outsourcing controversy, 121
Positivism, 134*f*, 156*n*6
Posting requirements, 78

Predator program, 45
Principal staff assistants, 68
Private sector
  arsenal model and, 27
  collaboration with, to manage complexity, 116–117
  conflicts of interest and, 61
  in contract model, 26t, 27
  expansion of, 116
  flexibility in, 3–4
  goals of, vs. government, 4
  increased use of, 119
  innovation in, vs. public, 40
  in Lead Systems Integrator model, 26t
  organizational complexity and, 34, 41
  in outsourcing model, 26t, 28–29
  as policy partner, 118–119
  as stakeholder, 112
  systems integration by, 59–61
  technical complexity and, 41
  viability, 36
  in weapon system manager model, 26t, 28
Process design, 95, 97, 97f
Process integration, 104–105
Process integrity, 95, 97, 97f, 104–106, 108t
Process models, 117
Products, vs. systems, 50
Product substitutability, 36
Program archetypes, 135t–136t
Program director competencies, 134, 138–139, 139f
Program diversity, 46
Program evaluation review technique (PERT), 11
Program executive officers (PEOs), 68, 72
Program frequency, 46
Program management
  in arsenal model, 26t
  authority, 100–101
  capability, 108f
  complex, 139
  complexity and, 45
  in contract model, 26t
  definition of, 25
  hierarchical structure in, 109n13
  keystone organizations in, 95
  in Lead Systems Integrator model, 26t
  life cycle phases and, 133–134, 134f
  lightweight, 96f
  modes of, 95
  organizational complexity and, 34
  outcome scores, 108t
  in outsourcing model, 26t
  premier, 96f
  theoretical frameworks, 98f
  in weapon system manager model, 26t
Programming code, in FCS, 92
Program requirements
  in arsenal model, 26t
  in contract model, 26t
  definition of, 25
  in Lead Systems Integrator model, 26t
  in outsourcing model, 26t
  in weapon system manager model, 26t
Project archetypes, 135t–136t
Project Categorization Framework (PCAT), 130–131, 131–133, 132t, 155
Project classification, 130–131, 131f
Project management
  accountability in, 101
  authority and, 100–101
  certifications, 131, 147–148
  challenges, 89–93, 90t, 91f, 93f
  changes and, 89
  classification in, 130–131, 131f
  complex, 129, 135t–136t
  complexity in, 89–90
  complexity of, 53
  complicated, 129–130, 135t–136t
  as continuum, 129–130
  distributed nature of, 90
  ecosystem efficiency and, 102–103
  ecosystem management and, 95, 96f, 108t
  ecosystem responsiveness and, 103–104
  experience and, 105–106
  experimentation and, 105–106
  FCS as, 92–93, 93f

Project management *(continued)*
 future of, 154–156
 life cycle phases and, 133–134, 134*f*
 managerial capabilities and, 93–98
 manager types in, 130
 nodes in, 129–130
 organizational barriers and, 100
 organization in, 99–100
 performance on management capabilities, 99*f*
 performance outcomes in selected, 90*t*
 pitfalls, 89–93, 90*t*, 91*f*, 93*f*
 process integration and, 104–105
 process integrity and, 104–106, 108*t*
 spectrum of, 129–130
 stakeholders in, 90
 structure, 99–100
 systems integration and, 52–53, 64*t*
 traditional, 129–130, 135*t*–136*t*, 156*n*2
 typology of projects in, 130–133, 131*f*, 131*t*, 132*t*
 uncertainty in, 90
Project manager career pathway, 145*f*
Project manager development, 148–149, 150*t*
Project manager life cycle development, 145–149, 145*f*, 146*f*, 150*t*
Project manager types, 130
Project typologies, 130–133, 131*f*, 131*t*, 132*t*
Pull method, 124–125
Purpose, unity of, 75
Push method, 125
Push-pull approach, 124–126

Raborn, William, 11
Radar, 10, 19
Recursiveness, 128
Regulatory environment, innovation and, 39
Relationships, interorganizational, 54
Repositories, data, 77
Requests for Proposals (RFPs), 117
Requirements
 in arsenal model, 26*t*
 completeness and, 84
 in contract model, 26*t*
 definition of, 25
 isolated defining of, 101
 in Lead Systems Integrator model, 26*t*
 in outsourcing model, 26*t*
 stability and, 84
 user, 110*n*16
 in weapon system manager model, 26*t*
Research, development, testing, and evaluation (RDT&E), funding as lever, 46
Resilience, complexity and, 2
Responsiveness, ecosystem, 103–104
Retention, of key individuals, 110*n*19
RFPs. *See* Requests for Proposals (RFPs)
Rickover, Hyman, 25
Risk
 assumptions, 18
 complexity and, 15
 governance and, 15
 "stress test" simulations for, 18
 testing, 18–19
 UBS bank losses and, 17
Risk management, 12, 15

Salomon Brothers, 18
Sapolsky, Harvey M., 6
SBIR. *See* Small Business Innovative Research (SBIR)
Schedule complexity, 133
Schrage, Michael, 5–6
Schriever, Bernard, 11
Scope certainty, 134*f*
Scope complexity, 134*f*
Scope uncertainty, 134*f*
Self-awareness, 143–144
Senior project manager level, 145–146
Sidewinder, 11
Simulation
 collaborative engineering and, 77*f*, 106, 110*n*15
 to manage complexity, 117
Singapore Defense Science and Technology Agency (DSTA), 148
Size, of enterprise, 69
Small Business Innovative Research (SBIR), 34–35

Software Engineering Institute, 55
SOS. *See* Systems-of-systems (SOS)
SOSA. *See* System-of-systems authority (SOSA)
SOSE. *See* System-of-systems engineer (SOSE)
Special Projects Office, 11
Sponsor role, executive, 123
ST. *See* Systems thinking (ST)
Stability
  as dynamic, 83
  engineering and, 84
Stakeholder(s)
  advocacy organizations as, 120
  complexity, 133
  external, 112
  in government, 112
  government as, 112
  identification of, 120
  increase in, 111
  informal organizations of, 113
  information flow, 113
  media as, 120
  in megaprojects, 90
  in networks, 111–112
  NGOs as, 120
  nonprofits as, 120
  persuasion of, 117
  policy formulation and, 115
  private contractors as, 112
  taxpayers as, 118
  think thanks as, 120
  universities as, 120
  variety of, 112
Stealth technology, 11
Stevens, Ted, 113
"Stovepipes," 70
Structure, in megaprojects, 99–100
Subprime mortgage crisis, 15–18
Substitutability, product, 36
Subsystems, complexity and, 33–34
Supplier networks, diverse, 60
System-of-systems engineer (SOSE), 72–75, 74*f*, 79–80
Systems
  definition of, 50
  longevity of, 54–55
  vs. products, 50

Systems integration
  academics and, 52
  adaptability in, 54–55
  conflicts of interest and, 56, 61
  contracting and, 61
  customer and, 53–54, 64*t*
  definition of, 50
  demand for, 51
  experience and, 56
  federally funded research and development centers in, 57, 62–63
  hallmarks of ideal, 51–57
  increasing importance of, 64–65
  by in-house laboratories, 58–59, 64*t*
  manufacturing expertise and, 64*t*
  organizational independence and, 59, 64*t*
  organizational longevity and, 64*t*
  organization types for, 57–64, 64*t*
  by prime contractors, 59–61
  project management and, 52–53, 64*t*
  skill in, 51
  technical awareness and, 51–52, 64*t*
  transparency in, 56
  trust and, 59
System-of-systems authority (SOSA), 72–75, 74*f*, 79–80
Systems-of-systems (SOS), 33
  analytical support, 73
  authority, 72–75, 74*f*
  "best," 67
  centralization and, 71
  characteristics, 70*f*
  collaborative environment and, 76–79, 77*f*
  culture and, 79–80
  definition of engineering, 71–72
  in DOD, 70–71
  environment for, 73
  independent, 70
  longevity of, 69
  network-centric engineering of, 71–76, 74*f*
  overlapping governance in, 69
  single missions and, 68
  unity of purpose and, 75
  wave planning for, 137*f*

Systems thinking (ST), 133–134, 134*f*, 139, 156*n*6

Taxpayers, as wild card, 118
Team complexity, 132–133
Team member, project manager as, 145
Technical awareness, systems integration and, 51–52, 64*t*
Technical complexity, 32, 33, 41, 47*n*4
Technical difficulty, 132
Technical direction
  in arsenal model, 26*t*
  in contract model, 26*t*
  definition of, 25
  in Lead Systems Integrator model, 26*t*
  in outsourcing model, 26*t*
  in weapon system manager model, 26*t*, 28
Technical execution
  in arsenal model, 26*t*
  in contract model, 26*t*
  definition of, 25
  in Lead Systems Integrator model, 26*t*
  in outsourcing model, 26*t*
  in weapon system manager model, 26*t*
Technical leadership, 15
Technology
  complexity and, 1
  FCS and, 93*f*
  gap, 4
  vs. political support, 25
  superiority in, 1
Tension, stability and, 83
Terrorism, 114
Testing
  governance and, 19
  risk management and, 18–19
Think tanks, as stakeholders, 120
"Third-best" solutions, 10
Thought leader role, 123, 126
Threats, FCS and, 93*f*
Three Steps to Victory (Watson-Watt), 10
Tizard, Henry, 10, 19
TPM. *See* Traditional project management (TPM)

Traditional project management (TPM), 129–130, 135*t*–136*t*, 156*n*2
Transparency, in systems integration, 56
Trust, in in-house laboratories, 59
Tyco, 13
Typologies, project, 130–133, 131*f*, 131*t*, 132*t*

UAVs. *See* Unmanned air vehicles (UAVs)
U-2 bomber, 11
UBS (investment bank), 16, 17–18
Uncertainty
  in megaprojects, 90
  scope, 134*f*
Underpinning knowledge, 141–142
Understanding, customer, 54, 64*t*
United States
  technological superiority of, 1
Unity, of purpose, 75
Universities, as stakeholders, 120
Unmanned air vehicles (UAVs), 33
Unmanned vehicles, 42
User requirements, 110*n*16

Value
  acquisitions and, 83
  operational, 83–84, 86
Viability, of industry base, 36
Virginia class submarine program, 37
Visibility, collaborative engineering and, 77–78, 77*f*

War on terror, 114
Watson-Watt, Robert, 10
Wave planning, 137*f*
Weapons acquisitions
  arsenal model of, 25, 26*t*, 27
  community of interest in, 24
  competition and, 36, 40–42
  complexity and, 40–42
  complexity in, 32–36
  complexity metrics and, 45
  contract model for, 26*t*, 27
  engineering and, 82
  environment for, 35
  evolutionary, 46

innovation in, 40–42
  Lead Systems Integrator model for, 26t, 29
  outsourcing model for, 26t, 28–29
  political will and, 24
  value and, 83
  weapon system manager model for, 26t, 27–28
Weapons systems
  complexity of, 32, 33–34
  as systems-of-systems, 33
Weapon system manager model, for acquisition, 26t, 27–28
WHOW matrix, 130–131, 131f, 132t
Will, political, 24
Wisdom, 143–144
Workforce capability, 42
WorldCom, 13

Zumwalt class destroyers, 33, 34, 37